8th International Conference on Mechanical, Automotive and Materials Engineering (CMAME 2020)

IOP Conference Series: Materials Science and Engineering
Volume 1194

Ho Chi Minh City, Vietnam
20 – 22 December 2020

ISBN: 978-1-7138-4653-6
ISSN: 1757-8981

Printed from e-media with permission by:

Curran Associates, Inc.
57 Morehouse Lane
Red Hook, NY 12571

CURRAN ASSOCIATES INC. proceedings.com

Institute of Physics
Dirac House, Temple Back
Bristol BS1 6BE UK

Phone: 44 1 17 929 7481
Fax: 44 1 17 920 0979

techtracking@iop.org

Additional copies of this publication are available from:

Curran Associates, Inc.
57 Morehouse Lane
Red Hook, NY 12571 USA
Phone: 845-758-0400
Fax: 845-758-2633
Email: curran@proceedings.com
Web: www.proceedings.com

TABLE OF CONTENTS

Preface

This proceeding contains the scholarly papers presented in the 8th International Conference on Mechanical, Automotive and Materials Engineering (CMAME 2020), which was held during December 25-27, 2020.

This conference invited professional experts from reputable international universities over the world to review the submissions. The papers in this proceeding have passed the vigorous review process and amended to the satisfaction of reviewers. Papers in the proceeding can be categorised as the following 2 topics: Material Performance Experiment and Analysis; Mechanical Design and Control Technology. CMAME2020 received more than 30 papers, 10 of them was accepted and published on the conference proceeding successfully. Meanwhile, this conference, attracts lots of scholars from Australia, China, Thailand, Russia, Vietnam, China, India, Jordan and Indonesia.

Because of the outbreak of COVID-19, CMAME2020 conference committees decide to change the onsite interactive academic conference to online remote conference. It's pity that we cannot have a face-face communication and discussion on advanced science and technology on mechanical, automotive and materials engineering this year for serious epidemic situation. But I would also like to thank the conference committee, the conference secretariat, keynote speakers and all participants for their hard work making this conference a success. Looking forward to the next meeting on 2021.

Professor John Mo
Royal Melbourne Institute of Technology, Australia
On behalf of CMAME2020 Conference Committees
December, 2020

IOP Conf. Series: Materials Science and Engineering 1194 (2021) 011001 doi:10.1088/1757-899X/1194/1/011001

CMAME2020 Conference Committees

Conference Committee Chairs

John Mo, Royal Melbourne Institute of Technology, Australia
Shane Xie, University of Leeds, UK

Program Committee Chairs

JING Xingjian, The Hong Kong Polytechnic University, Hong Kong
Yu-Liang Chen, National Defense University, Taiwan

International Technical Committees

Alokesh Pramanik , Curtin university, Australia
Dinesh Bhatia, The University of Nottingham Ningbo China
Eram Asghar, Politecnico di Milano, Italy
Gurpreet Singh, Mechanical Engineering, BBSB Engineering College, India
Hao Qin, Guangdong Institute of Intelligent Manufacturing, China
Jeefferie Abd Razak, Universiti Teknikal Malaysia Melaka, Malaysia
Kheng-Lim Goh, Newcastle University, UK
Meng Shaohua, Beijing Institute of Spacecraft Environment Engineering, China
Mohd Rizal Salleh, Universiti Teknikal Malaysia Melaka, Malaysia
Muhamad Arfauz A Rahman, Universiti Teknikal Malaysia Melaka, Malaysia
M. Chandrasekaran, Vels University, India
Nikolay K., POLYUYAN OVICH, Southern Federal University, Russia
Nur Izan Syahriah Binti Hussien, Universiti Teknikal Malaysia Melaka, Malaysia
Pooya Sareh, Imperial College London, UK
Prodip Das, Newcastle University, UK
Shamsul Jamel, Universiti Teknologi MARA, Malaysia
Thanakom Soontornchainacksaeng, King Mongkut's University of Technology North Bangkok, Thailand
Xiangsheng Gao, Beijing University of Technology, China
Zengtao Chen, University of Alberta, Canada

Peer review declaration

All papers published in this volume of IOP Conference Series: Materials Science and Engineering have been peer reviewed through processes administered by the Editors. Reviews were conducted by expert referees to the professional and scientific standards expected of a proceedings journal published by IOP Publishing.

- **Type of peer review:**
  Double-blind
- **Conference submission management system:**
  https://easychair.org/conferences/?conf=cmame2020
- **Number of submissions received:** 31

  Number of submissions sent for review: 21

- **Number of submissions accepted:**10

- **Acceptance Rate (Number of Submissions Accepted / Number of Submissions Received X 100):** 47.62%

- **Average number of reviews per paper:** 2 to 3

- **Total number of reviewers involved:** 15

- **Any additional info on review process:** We have first theme review on submissions...If the manuscripts are up to the first review standard, it will enter the second step----being reviewed by technical committees. As CMAME2020 technical committees, our reviewer will carefully check if this paper conforms to conference topic...The reviewers will discuss the strengths and weaknesses of manuscripts, contributions, and the potential impacts of the paper, while the innovation in the paper contents is necessary for CMAME acceptance. If the paper is weak...our reviewer will make suggestions for improvement and publication alternatives. We will contrast the similarity on submissions and published paper, if the similarity is up to 35%, the paper will be rejected for duplicate publication.

- **Contact person for queries:** John Mo, Royal Melbourne Institute of Technology, Australia, john.mo@rmit.edu.au

Corrosion Behavior of ASME SA-192 Steel Finned Tube by Wire Arc Spray in Temperature Control and High-Salt Environment

M Sangsuriyun[1]*, P Surin1[1] and K Eidhed[2]

[1] Department of Advanced Manufacturing Technology, Faculty of Engineering, Pathumwan Institute of Technology, Bangkok, Thailand
[2] Faculty of Engineering, King Mongkut's University of Technology North Bangkok, Bangkok, Thailand

*e-mail: montri.sang@npu.ac.th

Abstract. This paper proposes the corrosion behavior of ASME SA-192 steel finned tubes which were spray-coated with TH450 Chrome Nickel amorphous in a salt spray test chamber using 5% sodium chloride solution at a spraying speed of 12 mL/h according to The American Society of Testing and Materials (ASTM) B117 Standard. The salt spray test for 120 hours in the salt spray test in the chamber, the specimens that were exposed to moisture and salt fog in the surrounding air developed thin brown rust dots on the surface. The experimental results concluded that the specimens could withstand the corrosive environment for 5-15 years. The hardness test of coated specimens after the salt spray test for 480 hours showed that hardness and tensile strength of all specimens decreased due to surface damage from salt adsorption which resulted in blackish-brown rust and decreased tensile strength. Though, elongation of the coated specimens increased due to salt adsorption by the specimen surface, resulting in partial corrosion and rusting on the specimen surface. However, both parameters of the specimens exceed the ultimate tensile strength (UTS) standard test. Therefore, the spray coating with Chrome Nickel amorphous can increase the mechanical strength and corrosion resistance.
Keywords: Spray coating, ASME SA-192 steel, Mechanical properties, Cooling of finned tube

1. Introduction

The study of hardness improvement of ASME SA-192 steel finned tube using wire arc spraying process with TH450 Chrome Nickel amorphous, the cooling test showed no difference in the cooling rates between uncoated and coated specimens. The coated specimens will have a higher surface hardness which will increase the tensile stress of all the specimens and pass the Ultimate Tensile Strength (UTS) standard test. Therefore, spray coating a suitable technique for industrial applications that require higher hardness and can reduce maintenance costs. Arc Spraying is a thermal spraying technique that uses the heat released from electric arcing between two electrodes, in which the coating material must be in the form of conductive wire. Two wires are used as electrodes in an electrical circuit. These wires are fed from each side of the spray gun, and the end of each wire touches each other in front of the spray gun. Once electricity is applied through the wires, an electric arcing appears at the wire ends in front of the spray gun. The heat released from arcing increases and the wire ends are continuously heated to the melting point of the wire material. While high-pressure air or gas stream is applied at the wire ends from the back of the spray gun. This air or gas stream must have sufficient pressure to tear the molten wire material off the wire ends into molten aerosols which will then travel into the air or gas stream until these aerosols hit the surface of the specimen and harden and form a coating layer [1-3]. This technique is a suitable for industrial applications, including the production of new materials and repair work as an alternative to plating which requires higher hardness and corrosion resistance without affecting the toughness of the material [4-5]. A study [6] suggested that spray coating could prevent corrosion caused by salt or chlorine and the coating particles show high density with no crack but a few defects. According to the previous studies, the findings show that the specimen with the highest tensile strength does not always show the highest strength compared to the fatigue test [7].

In this study, we conducted the salt spray corrosion test of the finned tubes that were spray-coated with Chrome Nickel amorphous to simulate the potential exposure of products used in salt environments close to a coast or at sea, the mechanical properties and corrosion of the coated specimens were compared.

2. Experimental Procedure

2.1. Materials and specimen preparation

The materials used in this study were 30 fined tubes made of ASME SA-192 high-carbon coated steel by wire arc spraying process. These finned tubes consist of 0.6-0.18% C, 0.27-0.63% Mn, 0.25% Si, 0.035% P and 0.035% S. The tube has a diameter of 0.38 mm, the wall thickness of 3.2 mm, height and thickness of fin are 16 ± 1 mm and 1 mm, respectively. These specimens in this study were prepared according to the standard finned tube production [7].

2.2. TH450 Chrome Nickel amorphous coated wire and spray parameters

In wire arc spray, chemical properties of the wire used for TH450 Chrome Nickel amorphous coating consist of Cr 30%, Nb 10%, B 4%, Ni 10%, Fe Bal and other substances <50% by weight. Besides, the spray in this research determines the parameters as follows diameter 1/16" (1.6 mm), air pressure 80-100 psi, voltage 28-32 psi, amperage 100-300 and standoff 4-7" (10-17 cm) [8].

2.3. Wire arc spray coating

Before spray coating, rust stains on all 30 specimens must be removed by sandblasting and then wiped clean. After that the specimens were spray-coated with ASME SA-192 high-steel carbon using TH450 Chrome Nickel amorphous by wire arc spraying process using Thermion Auto Arc AVD 3350, AHB-20A/D, air compressor at 20 HP, 15 kW, air 6 bar, 37.5 voltage and 19.6 A. All specimens were spray-coated, which default values of spraying as the following parameters: air compressor, 20 HP/15 KW, model AH 20A 8 bar, voltage 37.5 V, electric current flowing through a conductor 19.6 A and temperature 926.67°C or 1,700°F .

2.4. Corrosion test

Corrosion testing procedure of the coated specimens with TH450 Chrome Nickel amorphous using Salt Spray Tester TU380 is as follows: all 30 specimens were placed in the chamber which was designed and controlled according to ASTM B117 standard. The corrosion test was performed at 38°C by using 5% NaCl at a flow rate of 12 mL/h. The salt fog accumulated in both volumetric cylinders for at least 80 mm^3, in which the first cylinder was kept near the salt spray nozzle and another cylinder was placed farthest from the sprayer. Visual inspection of specimens was then performed and compared with the microscopic examination results at 100 μm magnification on the 24, 120, 240, 360 and 480 hours [9].

2.5. Micro- and macrostructural analysis

The micro- and macrostructure analysis of all seven specimens was performed using an inverted microscope model Zeiss Vert A1 in order to investigate the corrosion caused by the salt spray test.

2.6. Hardness test using Micro Vickers Hardness Tester

After the salt fog was sprayed on the surface coated specimen with TH450 Chrome Nickel amorphous in the Salt Spray Tester TU380 chamber and left for 480 hours, the hardness test of all seven specimens was performed with the design parameters as [7]. Before the hardness test, the specimen surface was polished using 180 - 1200 grit sandpaper and then finished using felt cloth and 0.3 – 0.1 μm alumina powder. The specimens were then etched with a mixture of nitric acid and alcohol (10:90) to expose the microstructure of the specimens. After these pretreatment steps, the hardness test of all specimens was performed with 0.98 kg compression load using Micro Vickers Hardness Tester model HV-1000. The results were expressed as $HV_{0.5}$.

2.7. Tensile strength test

The tensile test of specimens was performed at Disp of 0.500 mm/min, 1.00 kN/s or 10 MPa/s load and 0.0500 mm/s extension speed in triplicate per treatment to compare between the specimens with and without salt spray test. Then the specimens are tested for tensile strength in order to compare with the standard ultimate tensile strength (UTS) that is higher than 350 MPa [10].

3. Results

3.1. Salt spray corrosion test of coated specimens with Chrome Nickel amorphous

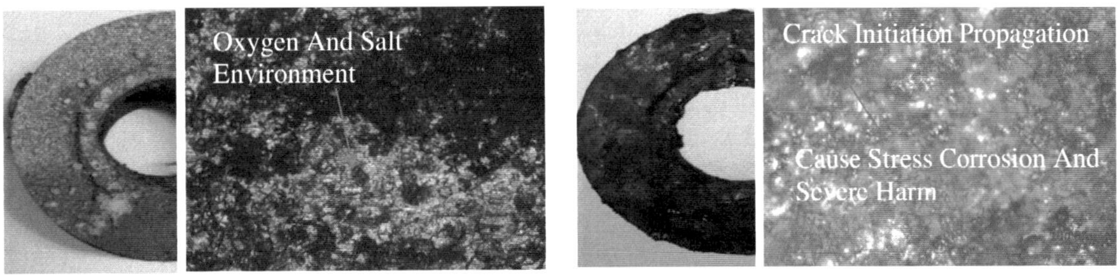

(a) Corrosion characteristics, 24 hours. (b) Accelerated corrosion test, 480 hours.

Figure 1. Corrosion characteristics after accelerated corrosion test.

The corrosion test using Salt Spray Tester TU380, the rusting or oxidation reaction $Fe + H_2O + O_2 = Fe_2O_3H_2O$ [11] occurred on the surface coated specimen with Chrome Nickel amorphous on the first day or within 24 hours as shown in figure 1 (a).

On 480 hours, the visual inspection of coated specimens showed that there were blackish-brown rust layers all over the specimen surface and the microscopic examination at 100 μm magnification showed that there was only 10% of Chrome Nickel amorphous coating layer left on the specimen surface as shown in figure 1 (b). The corrosion appeared as wells and rust layers.

In conclusion, the results showed that after the coated specimens with Chrome Nickel amorphous was sprayed with salt fog in Salt Spray Tester in the chamber for 120 hours, corrosion occurred did not severely affect the strength and specimens. So, the specimens can withstand corrosion in normal environments for 5-15 years [12]. However, when continuing the test, the rusting of the specimen in the chamber on 480 hours caused by the oxidation due to moisture and salt in the surrounding air, which would affect the safety.

3.2. Compare the hardness of the ASME SA-192 steel finned tubes using Nickel Chromium amorphous coating with salt spray testing.

Figure 2 (a) shows the test points for hardness test using Micro Vickers Hardness Tester comparing the hardness after the salt spray test (NiC-S) of coated specimens with Chrome Nickel amorphous. As shown in figure 2 (b), the hardness at point 1-3 and point 8-10 were similar and were in the range of 400-650HV because these areas were coated with Chrome Nickel amorphous and all seven specimens were spray-coated simultaneously. However, the heat-affected zone (HAZ) was observed between the coating layer, and the tuned point of the specimen was affected by the heat resulting in higher hardness compared to the heat-affected zone, and hardness at point 1-7 of all specimens decreased to 131-155$HV_{0.5}$.

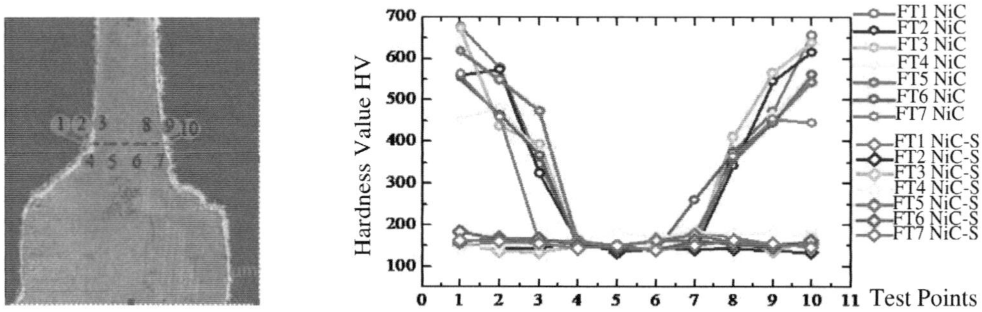

(a) Test points (b) Comparison of the hardness of specimens

Figure 2. The hardness of specimens that are coated with TH450 Chrome Nickel amorphous, without salt spray test (NiC) and with salt spray test (NiC-S) using Salt Spray Tester TU380.

For NiC-S method, after the salt spray test in Salt Spray Tester TH380, the hardness at the coated points significantly decreased. The average hardness at point 1-10 was in the range of 150-190 $HV_{0.5}$ as shown in figure 2 (b) which was consistent with the microscopic examination results showing saline adsorption and blackish-brown rust on specimen surface on 24 hours which increased continually until 480 hours. Corrosion appeared as rust layers covering a wide area of the specimen surface and resulted in decreased in hardness.

3.3. Compare the tensile properties of the ASME SA-192 steel finned tubes using TH450 Nickel Chromium amorphous coating with and without salt spray testing using Salt Spray Tester TU380

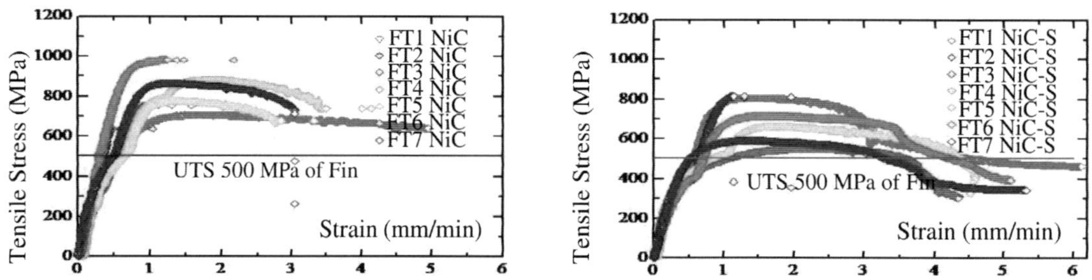

(a) (b)

Figure 3. Comparison of the tensile strength of the coated specimens with TH450 Chrome Nickel amorphous without (NiC) and with salt spray test (NiC-S) using Salt Spray Tester TU380.

The comparison of the tensile strength of the coated specimens with Chrome Nickel amorphous, the specimens without salt spray as shown in figure 3 (a) showed high tensile strength and exceed the UTS standard test, while the coated specimens with Chrome Nickel amorphous and sprayed with salt fog in the Salt Spray Tester TU380 (NiC-S) showed a decrease in tensile strength as shown in figure 3 (b). The accelerated rusting using sodium chloride solution with NiC-S method showed that there were two types of rust, reddish-brown and blackish-brown rush dots and cracks and crack propagation in the blackish-brown corrosion holes were also found. A study [12] suggested that as tensile strength decreased, elongation of specimen increased compared to that of uncoated specimens because Chrome Nickel amorphous coating layer was only partially damaged by rust. The experimental found that all specimens showed high tensile strength and exceed the UTS standard test.

IOP Conf. Series: Materials Science and Engineering 1194 (2021) 012001 doi:10.1088/1757-899X/1194/1/012001

4. Conclusions and Recommendations

- In this study, the salt spray test of ASME SA-192 steel finned tubes that are coated with TH450 Chrome Nickel amorphous was studied using Salt Spray Tester TU380. The results showed that arc spray coating with TH450 Chrome Nickel amorphous could increase the corrosion resistance against salt spray. The corrosion was not severe and did not affect the strength of specimens. This coating technique can extend the useful life of specimens by 5-15 years in normal environment according to ASTM B117 standard. The coating can be done many times, resulting in reduced maintenance costs. And this method should be used in other materials.
- The hardness of the specimens sprayed with salt fog using the Salt Spray Tester TU380 decreased after 480 hours of testing. Besides, the specimens showed a decrease in tensile strength. However, coating with Chrome Nickel amorphous could protect the specimen surface and result in partial damage and higher elongation compared to uncoated specimens. The results also showed that all specimens showed high tensile strength and exceeded the UTS standard test.

References

[1] Kelkar M, and Heberlein J 2002 Wire-arc spray modeling. *J. Plasma chemistry and plasma processing*. **22** 1-25
[2] Ducos M and Durand J P 2001 Thermal coatings in Europe: a business perspective. *J. Thermal Spray Technology*, **10** 407
[3] Hutchings I and Shipway P 1992 *Tribology: friction and wear of engineering materials* vol 2nd, ed Butterworth-Heinemann 13 187
[4] Zahrani E M and Alfantazi A M 2012 Molten salt induced corrosion of Inconel 625 superalloy in $PbSO_4$–Pb_3O_4–$PbCl_2$–Fe_2O_3–ZnO environment. *J. Corrosion Science*. **65** 340-359
[5] Rigney R W, Grubowski A, McCaw R and Scandell K 2001 *Component repair and chrome plating replacement with new thermal spray in the united states navy: successes and the future*. Materials Characterization, February-March.
[6] Qin E, Yin S, Ji H, Huang Q, Liu Z and Wu S 2017 Hot corrosion behavior of arc-sprayed highly dense NiCr-based coatings in chloride salt deposit. *J. Thermal Spray Technology*. **26** 787-797
[7] Sangsuriyun M, Surin P and Eidhed K 2006 Optimization of high-frequency resistance welding process using mechanical property of finned tube SA-192 steel. *J. Engineering and Applied Sciences*. **15** 607-612
[8] Basumatary J, Wang H, Watson S,Powrie H E G 2005 The corrosion of nickel-aluminium bronze in seawater [in A Century of Tafel's Equation: A Commemorative Issue of Corrosion Science] *J. Corrosion Science*. **47** 3336-3367
[9] Wan Y, Macha E N and Kelly R G 2012 Modification of ASTM B117 salt spray corrosion test and its correlation to field measurements of silver corrosion. *J. Science and Engineering*. **68** 036001-1-036001-10
[10] ASTM D and others 1995 Standard test method for tensile properties of thin plastic sheeting. *J. American Society for Testing and Materials*. ed West Conshohocken 194-202
[11] Hua Y, Barker R and Neville A 2015 Understanding the influence of SO_2 and O_2 on the corrosion of carbon steel in water-saturated supercritical CO_2. *J. Corrosion*. **71** 667-683
[12] Van Boven G, Chen W, Rogge R and Sutherby R 2007 The Effect of Residual Stress on Pitting and Stress Corrosion Cracking of High Pressure Natural Gas Pipelines. *J. Acta mater*. **55** 29-43

Neural Network Multistage Method for The Forecasting Electrical Insulating Materials Characteristics

Poluyanovich Nikolai Konstantinovich, Azarov Nikolai Vasilievich, Ogrenichev Alexander Vasilievich and Dubyago Marina Nikolaevna

Southern Federal University, Institute of Radio Engineering System and Control, Department of Electrical Engineering and Mechatronics, Russia, Taganrog, Chekhov st., 2, 347922

Email address: nkpoluyanovich@sfedu.ru

Abstract. The solution to the problem of non-destructive monitoring of thermal processes in insulating materials is shown. The main configuration parameters of the artificial neural network are determined. A mathematical model and architecture of a neural network for the forecasting thermal fluctuation processes are developed. The architecture of a neural network with a consistently distributed structure is synthesized. A forecasting technique is proposed and functions are obtained for assessing the long-term permissible heating values of the cable current-carrying core insulating material. Temperature monitoring will allow to monitor both the condition and the throughput of the lines.

1. Review of systems for the forecasting thermal processes

Insulation materials (IM) during operation are subject to thermal, electrical, chemical, mechanical and other types of influences [1]. The temperature rise increases the likelihood of a decrease in the IM electrical strength, which further promotes their breakdown. Therefore, the main task of the operating personnel is to control and maintain the IM electrical insulating properties at a level that excludes their emergency failure [2, 3]. The direct destruction of the polymer is explained by cracks formation in the stress concentration places and the initiation and development of cracks is considered as a consequence of the kinetic process of thermofluctuation breaking of bonds [1]. It was found that the source of breaks in the structure of an insulating polymer is thermal fluctuation processes, which sharply increase the kinetic energy of individual atoms vibrating around the equilibrium position, as a result of which the chemical bond of the polymer backbone sometimes breaks [2-4]. Chemical degradation will cause changes in the mechanical properties of polymer insulation [5]. A thermodynamic approach is needed in the study of the processes of the insulating materials destruction. The IM low thermal conductivity of the power cable line (PCL) in Fig. 1 leads to a high temperature gradient, therefore, the temperature of the most heated part in the PCL section (near the core θc) significantly differs from the IM temperature on the casing ($\theta p.c$), and in transient modes this the difference can increase several times [3, 6].
The object of the research is APvPu g-1x240 / 25-10 [3] IM with cross-linked polyethylene (XLPE) insulation. The CL thermal field is described by the differential equation of thermal conductivity (1), which in the general case determines the change in the temperature distribution in the volume of the object under consideration depending on time [7]. In [8], another solution was obtained that allows to determine the cable surface temperature using an integral exponential function. The proposal to calculate the temperature on the cable surface using an exponential integral was implemented in [9], where the temperature difference between the cable surface and the soil surface was calculated using the theory

[8], and the temperature difference between the core and the outer casing, as in [10], was calculated using a thermal equivalent circuit. But the calculation of its parameters in [12] is performed using the Van Wormer method [11]. Since the heat wave from the core and the screen does not immediately reach the cable surface, a certain coefficient a(t) ("attainment factor") was introduced in [9] for matching between the two calculated regions. To calculate the core and screen temperatures, considering the constantly changing core current and screen temperature, the authors of [12, 13] presented them in the form of a stepped graph. The influence of each current jump and screen temperature was considered. The review has shown that most of the methods have low forecasting accuracy, are difficult to forecast the thermal processes in PCL, and are not capable of operating in real time; therefore, it is necessary to search for a promising non-destructive method.

1.1 Influence of the inclusions presence on the thermal processes

There are inclusions in the IM and they arise during operation (Fig. 1) [7], which affect its heating and deterioration. Let's simulate the IM resource (service life) depending on the temperature, formula (4).

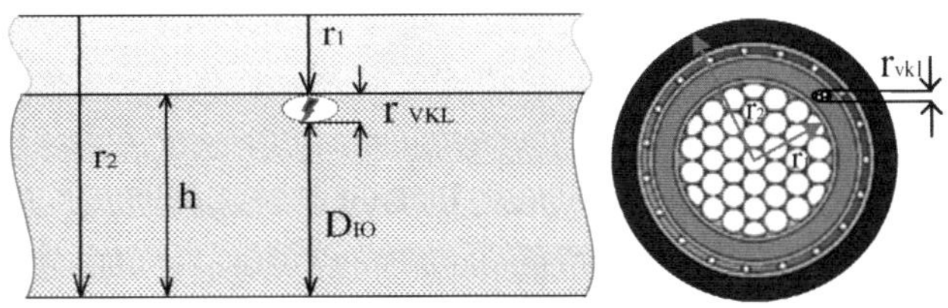

Figure 1. The investigated PCL insulating material

The total electric field intensity in the IM for different sizes of the inclusion (r_{VKL1}, r_{VKL2}, r_{VKL3}) is equal to

$$E = E_V + E_{OI},\ npu\ r_{VKL} = \text{var}\,iable \quad (1)$$

Considering that the intensity on the inclusion for the radius r_{INC1}, r_{INC2}, r_{INC3}

$$E_V = UC_I \Big/ 2r_{VKL}C_V log\left(\frac{h}{2r_{VKL}}\right) \quad (2)$$

Intensity on the residual insulation layer for the radius r_{vkl1}, r_{vkl2}, r_{vkl3}

$$E_{OI} = U\frac{\dfrac{C_{V1}C_{OI1}}{C_{V1}+C_{OI1}}}{4\pi\varepsilon_0\varepsilon_{VKL}\left(h-2r_{VKL1}\right)} \Bigg/ log\left(\frac{h}{D_{OI1}}\right)\left(h-r_{VKL1}\right) \quad (3)$$

When calculating the IM residual service life, it was considered that one monomer contains two CC bonds and four CH bonds. The breaking energies of one bond were taken, respectively, W_{CC} = 84 [kJ/(mol)] and W_{CH} = 99 [kJ/(mol)].

The calculation of the CL insulating material service life is performed according to the formula

$$\tau = \tau_0 exp\left(\frac{Wcp - \chi E}{RT}\right),\ \text{where}\ Wcp = \frac{2W_{CC}+4W_{CH}}{6}. \quad (4)$$

where χ – structural parameter [kJ mm / mol kV] = 0,447; E – field intensity.

Forecasting results of the PCL IM residual life are shown in Figure 2.

The dependence of the IM resource on the inclusion size is obvious (Fig. 2), and, consequently, the energy loss in the IM and its heating. Therefore, the problem of temperature monitoring is relevant, considering the inclusions in the IM [3].

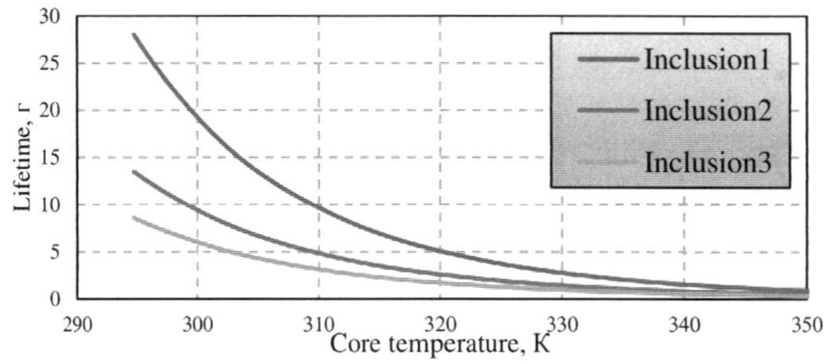

Figure 2. Forecasting the IM residual life

2. Experimental researches

In order to identify the defects occurrence during the cable operation, conditions for its artificial aging were created. For this, at the first stage, the samples of the cable under study were kept at the same temperature, but with different time durations in a heating cabinet. After heat treatment, the cables were subjected to current loading, which resulted in the cable insulating material heating. Graphs of temperature changes (experimental and forecasted) of one of the investigated cables before and after aging (stages 1 and 2) are shown in Fig. 3.

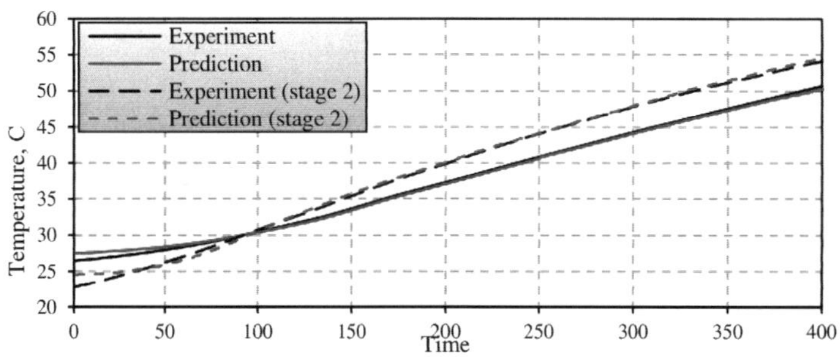

Figure 3. Experimental and forecasted temperature

Further the comparing of the temperature difference was made. The research results for "new" (training sample) and "aged" (control sample) cable samples are summarized in Table 1.

Table 1. Parameters comparison

Parameter		№ cable sample		
		1	3	4
Temperature difference ΔT	New	1.73	2.8	2.3
	Aged	1.2	1.46	1.6
Error, %	New	4.2%	5.2%	3.8%
	Aged	3.8%	2.3%	3.1%

Analysis of the results showed that the longer the aging time, the greater the temperature difference between the original and the aged sample. With increasing aging time, the temperature difference ΔT increases. The temperature difference (ΔT) allows to determine the insulating material residual life [7]. At the same time, the forecast error of the neural network decreases, that is, the accuracy of forecasting the cable core temperature increases.

We believe that this is due to the loss of the material electrical insulating properties due to the accumulation of the destroyed structure fragments, containing, in ever greater quantities, the inclusion of pure carbon and other conductive inclusions. The reaction of incomplete oxidation of polyethylene with the formation of carbon:

$$\left[-CH-CH2\right]n+nO_2\rightarrow 2nC+2nH_2O \tag{5}$$

Thus, the dependence of the increase in the temperature difference ΔT on the deterioration degree of the material electrical insulating properties has been revealed and investigated, which makes it possible to forecast its resource.

2.1 Multipurpose optimization of NN forecasting method

The result of the carried out multipurpose optimization showed that the efficiency of the NN methods using is 1.6-1.9 times higher than that of traditional methods [14]. The solution to the forecasting problem using models based on artificial neural networks is proposed [19-20].

2.2 Mathematical description of the algorithm

In the mode of dynamically changing load of the power cable, it is necessary to forecast its temperature regime. A transformation algorithm has been developed that allows from the members of the time series (X (n)) of the input parameters of the PCL: θp.c (1 ... n) is the temperature of the protective casing, θe (1 ... n) is the environment temperature, Ic (1 ... n) - current in the PCL core to form a "window", which is a training set for building a forecast. The window is a time interval containing a set of values (θp.c (1 ... n), Ic (1 ... n), θe (1 ... n)), which are used to form a training example.

A serial two-network architecture of a neural network with a sequentially distributed structure is synthesized. The first network forecasts the PCL surface temperature, the second forecasts the core temperature. The mathematical model of the NN is given below.

$$\begin{cases} g_i = f\left(\sum_{i=1}^{n} v_{ij}x_i + Q_j\right), y_j = f\left(\sum_{j=1}^{h} w_{jk}g_j + T_k\right); \\ w_{jk}(t+1) = w_{jk}(t) + a\left(\partial E/\partial w_{jk}(t)\right); \\ T_1(t+1) = T_k(t) + a\left(\partial E/\partial T_k(t)\right); \\ \dfrac{mN}{1+\log_2^N} \le Lw \le \left(\dfrac{N}{m}+1\right)(n+m+1)+m; L = \dfrac{Lw}{n+m} \end{cases} \quad (6)$$

2.3 Consideration of magnetic interference with dynamic current change

The process of the core current changing is dynamic, although time is not explicitly included in the dynamic law of the current in the core (I_c), written in differential or recurrent form, therefore the process is called stationary.

The diagrams of the surface experimental and forecasted temperature and the temperature of the PCL central core are presented (Fig. 4, b) depending on various values of the current in the core (Fig. 4, a) and considering the interference from the neighboring cable.

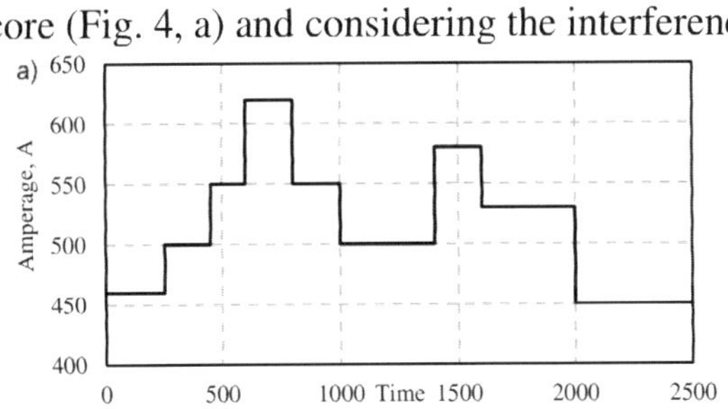

Figure 4 a - dynamics of current change

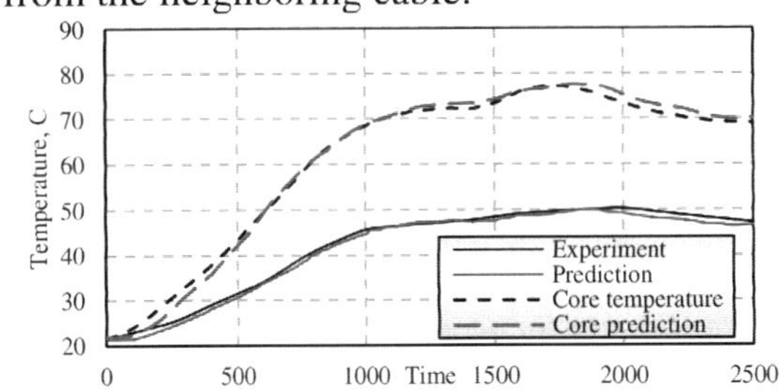

Figure 4 b - cable temperature

The results of the temperatures comparison and the forecast error are shown in Table 2.

Table 2. NN forecast considering account the influence of magnetic interference

Magnetic interference	Average temperature / forecast error	
	Experimental, ° C / average, ° C	Forecasted, °C / t, %
No interference	61,2 °C /1,06 °C	61,47 °C /1,73%
With interference	61,2 °C /1.62 °C	60.68 °C /2.64%

Thus. The results of the NN forecast showed that for cables with XLPE insulation it is possible to forecast the current-carrying core temperature. Considering the influence of magnetic interference on the cable showed that the presence of magnetic interference on the cable (taking them into account as an input parameter of the NN) increases the error by 1.3 degrees (or 0.8%).

Conclusion

A set of studies was carried out to improve the accuracy of the thermal processes forecasting in the cable lines insulating materials, considering changes in the current load of the line and external conditions of heat removal, air temperature and magnetic field strength. The comparisons of the forecasted values with the actual ones allow us to speak about the adequacy of the selected network model and its applicability in practice. The developed methods and algorithms are elements of an integrated diagnostic system and are aimed at efficient and reliable power supply to consumers. The use of the developed neural network system for the power cable temperature forecasting contributes to the planning of operating modes of power supply networks for consumers.

References

[1] Ukil, A., Braendle H., Krippner P. 2012 *Sensors Journal,* Vol. 12 (IEEE) No. 5. pp. 885–892.
[2] Dubyago M.N., Poluyanovich N.K. (2018) *Endorsed Transactions on Energy Web and Information Technologies* (EAI) 18(16) **e3.**
[3] Dubyago M.N., Polyuyanovich N.K., Burkov D.V. 2019 *Energy Web* Vol.6 (EAI) 19(24) e3
[4] Dubyago M.N. and Poluyanovich N.K. 2017 *Materials Science and Engineering 229* Vol.5 *(IOP)* Issue 16. e3.
[5] Sugimoto, M. and oth. 2013 Product analysis for polyethylene degradation by radiation and thermal ageing *Radiation Physics and Chemistry* Vol.82 (Elsevier) № 1. pp. 69–73.
[6] IEC Standard 60287–1–1. Electric Cables – Calculation of the Current Rating. Part 1: Current Rating Equations (100% Load Factor) and Calculation of Losses – Section 1: General. – 2006.
[7] Dubyago M.N. and Poluyanovich N.K. 2019 *International Journal of Materials* Vol. 7 (Mechanics and Manufacturing), no. 1, pp. 1-6.
[8] Li, H.J. 2005 Estimation of soil thermal parameters from surface temperature of underground cables and prediction of cable rating *Generation Transmission and Distribution* Vol. 152 (IEEE) № 6.pp. 849–854.
[9] Dubyago M.N., Polyuyanovich N.K 2018 *Endorsed Transactions* on *Energy Web* (EAI) 5(16): 153815.
[10] Millar, R.J., Lehtonen M. 2005 *15th Power Systems Computation Conference* Vol. (Belgium: Liege) Session 39. Paper 4. pp. 1388–1395.
[11] Van Wormer, F.H. 1955 *Power Apparatus and Systems* Vol. 74 (AIEE) Issue 3. pp. 277–281.
[12] Prime, J.B., Valdes J.G 1981 *Power Apparatus and Systems.* Vol. PAS–100 (IEEE) № 1 pp. 211–219.
[13] Mazzanti, G. 2009 *Dielectrics and Electrical Insulation* Vol. 16. (IEEE) № 4. pp.1168–1179.
[14] Akhmedova S., Semenkin E. 2014 *Congress on Evolutionary Computation* (IEEE) pp. 867–872.
[15] Chen X., Xu Y., Cao X. 2014 *Engineering, Computer Science* (IEEE) № 4. pp. 198-205.
[16] A. Shimada 2013 *Dielectrics and Electrical Insulation* Vol. 20 (IEEE) № 6. pp. 2107–2116.
[17] Lee K. Y., Park J. H. 1993 *Power Systems.* (IEEE TransactionsVol. 7) No 1. pp.1030 - 1038.
[18] Meldorf M., Kilter J., Pajo R. 2007 *CIGRE Regonal Meeting*, (Estonia: Tallinn) pp.145-150.
[19] Terracciano M., Purushothaman S. 2012 *Power Delivery* Vol. 27 (IEEE). № 3 pp. 1423 – 1431.
[20] Belyavskiy G., Puchkov E. 2017 Vol.115 (International J. of Pure and Applied Mathematics.) No.4 pp. 883-893.

Acknowledgments

This work was supported by the RFBR grant No. 18-08-00473 on the theme: “Development and research of methods for optimizing and controlling the processes of energy conversion in power plants of a complex type, including those converting a continuous flow of medium”.

The Ideal Temperature Setting of Polymer Blends: *Investigational Characterization Effects of Color Matching*

J. Al Sadi, A. Hawary

Jadara University, Irbid, Jordan

Email: jamal.alsadi@ontariotechu.net, j.alsadi@jadara.edu.jo

Abstract. The goal of this work, to study the characteristics of a mixture of the two Lexan polycarbonates, having different (MFI) the characterizations were run in two stages. The first stage: The two-polycarbonate resins of different weight percentages and MFI were melt-blended using (Coperion) a Co-rotating twin-screw extruder (SB). This grade rheologically characterized using the rotational rheometer and MFI tester. The same material was included; the same composition was blended in steps of eleven in a Thermo Haake Mini Lab II twin-screw micro compounder (ML). Further experimental design dictated a sample without pigment and additive (WOP) consisting of only PC resin, a sample with PC resin mixed with pigment and additives (WP) to study the characterization of polycarbonate formulation at different temperatures were also analyzed. They were rheologically characterized using the rotational Rheometer, MFI tester, Particle size analyzer and Scanning electron Microscopy (SEM). This work was mainly focused on examining the viscosity at the mid (Centre) - temperature =255°C. The focus was extended to the polymer grade to extract the impact of the rheological and dispersion characteristic at 255°C to study their correlations in the viscosity data and their bearing effects on color matching.

1. Introduction

Producing plastic with a marketable color requires the addition of one or more resins; however, achieving the correct color in the first attempt is a challenge. The color properties of the polymer are directly affected by formulation for various ratios of PC resins. The viscosity for a certain ratio may vary due to the effect of formulation such as (PCs weight percent blends, pigments, and additives), and processing parameters. A few studies have been carried out by various researchers regarding the effects of processing parameters by the dynamic mixing in a screw extrusion during polymers' compounding [1, 2]. In order to control the mentioned points, a system of additives and resins can be formulated to modify viscosity and improve thermal stability to have better dispersion [3]. Researchers have studied the importance and effects of adding the pigment to the base resin; especially, it has been shown that incorporating additives into polymeric materials during production often unpredictably affects rheological and optical properties [4-6]. However, rheological properties are an essential link between the processing steps and the final performance of the product [7-8].

Polymer blending is an important field of polymer science and has been reviewed by many scientists. The PC/PBT blends are transparent in the melt state, and most transparent in the solid-state are partially miscible blends, as was shown by Sanchez et al. [8]. In their work, Liang and Gupta (2000) studied the rheological properties of a recycled PC blended with virgin PC; they stated that separated PC could be added to pure PC up to the 15% level without significantly varying the properties of pure PC [9]. Khan et al. developed in their study characterizing alloys of ABS and PC through blending these two materials [10]. Lee S. et al. studied PC/polyester blends' rheological behavior and their phase behaviour [11]. However, they found that the blends do not follow the rule of mixtures which is common in all studies.

The particle sizes, fillers (pigments and additives), and processing parameters like temperature affect the rheology and the pigmented polymer's color. The significance of the processing variables was correlated with the rheological result [12].

Difficulties with the dispersion of pigments or obtaining a uniform blend, can be overcome by reducing the resin viscosity and extending the mixing time [13]. A system of additives and resins can be formulated to modify viscosity, increase mechanical prosperities, improve thermal stability, or improve wear performance [14]. Many researchers have studied the importance and effects of adding the pigment to the base resin, especially since it has been shown that incorporating additives into polymeric materials during production often affects rheological, mechanical, and optical properties unpredictably [15, 16,17]. Rheology, as a conception, applies to all types of materials (solid, semi-solid and fluid) like polymers and their composites. Rheological properties are an important link between the processing steps and the product's final performance [18, 19].

This study included additional literature reviews, investigation for the materials' rheological characterization, and Melt flow adjustment (MFA). Haake mixer and viscosity measurements [20, 21, 22, 23, 24]. More recent studies have indicated that the effective processing conditions were dependent on the materials used or blending different materials (resins, additives, pigments, and fillers) to improve the rheological characteristics [25, 26, 27, 28]. The effect of particle sizes, parameters like temperature, pressure, and amount of additive can be determined and used for improving processing conditions [29, 30].

Many researchers have studied the degradation pathway for the polymer. The degradation Photo-oxidative of weathered plastic constituents is due to the combined effect of UV light (295nm to 400nm), which is existing in the solar spectra, moisture, and heat [31]. Most polymers' primary degradation pathway is measured to be their reaction with atmospheric oxygen, consequential resulting in chain cleavage, crosslinking, and discoloration. Although photodegradation of polycarbonate (PC) resin extensively been measured [32-34], more investigation was continuing about the effect chemistry involved. When manufacturing plastics, e.g., bisphenol, a polycarbonate (PC), are used in outdoor applications, oxygen, sunlight, humidity, and can be the source of degradation of the polymer leading finally to mechanical failure and loss of optical and aesthetical properties [35]. The instrument settings were as follows: color space – CIE Lab; measurement mode; observer angle – 10° and D65 illuminant (light source) [36]. Taking the above results into account makes sense that the addition of colorant will minimize the viscosity, and the absorbance mechanisms here for degradation and yellowing will be decreased as was seen for samples subjected to an earlier study [37-38].

This work *was* mainly *focused* on examining the viscosity at the mid (Centre) - temperature =255°C. The focus was extended to the polymer grade to extract the impact of the rheological and dispersion characteristic at 255°C to study their correlations in the viscosity data and their bearing effects on color matching.

2. Experimental procedure and materials

This research investigated two PC resins grades referred to as R1 and R2 in the thesis, each having a different melt flow index (MFI). One had an MFI of 25, and the other had an MFI of 6.5g/10min, respectively, here forth. The resins are manufactured by General Electric (GE) and traded under the name of Lexan. Four different color pigments, black, white, red and yellow, were used. Also, three additives were used called F1, F2, and F3 in this work. One was a stabilizer, one a light stabilizer, and the third offers weather-resistant properties. Materials were extruded in an intermeshing, 25.5 mm, Coperion twin co-rotating screw extruder(SB). The total weight of the color additives (pigment and additive) was 0.86%. The two PC resins, R1 and R2, were used in a ratio of 30 and 70 wt. %, respectively. The additives and pigments were mixed with the resins at a 100:0.86 ratio and were blended by a super floater. [39-40] Table 1 shows the formulation used.

Table 1. The composition of compounding material (g3)

Ingredients	Material Name	PPH
R1	Bisphenol A (BPA)	30
R2	Bisphenol A (BPA)	70
F 1,F2,F3	Additives	0.3
W,B,R,Y	Pigments	0.56

To study the effects of the blending of two resins, a Thermo Haake Minilab II twin-screw micro compounder (ML) was used to prepare PC formulations with both resins individually as well as by varying the composition of the resins in steps of ten for a total of eleven blends. The concentration ratios between the two polycarbonate resins used in %R1/%R2 were 100%/0%), 90%/10%, 80%/20% and 0%/100%.

3. Results & Discussion

3.1 Rheological behaviour at 255°C.

The PC blends were rheologically characterized at 255°C using the Ares rotational rheometer. Results of complex viscosity at 255°C vs. frequency for the blends without additives (WOA) are presented in **Figure 1.a**. Blend 100% R1 exhibited the lowest viscosity with a solid-like behaviour at the low frequencies compared to blend 0% R2, while both presented shear thinning behaviour at the higher frequencies. With the addition of R2 into the R1 resin, a viscosity of the blends increased substantially. The dynamic responses of R1/R2 blends do not appear to follow the "rule of mixing," which expects a variation of the parameter to be in proportion to the amounts of constituent elements of the blend. The increase in viscosity with the addition of R2 may be due to the solvation of the highly entangled structure of R2 molecules in R1 molecules. Similarly, **Figure 1.b** presents the viscosity of blends where the pigments and additives (WA) are added. Resin R2 (100%) shows a higher viscosity and exhibits a strong shear thinning effect while Resin R1 (100%) shows a lower viscosity in comparison to 100% R2 and the other blends. In general, the blends' viscosity decreased in the presence of pigments and additives by approximately 10-20%.

(a)

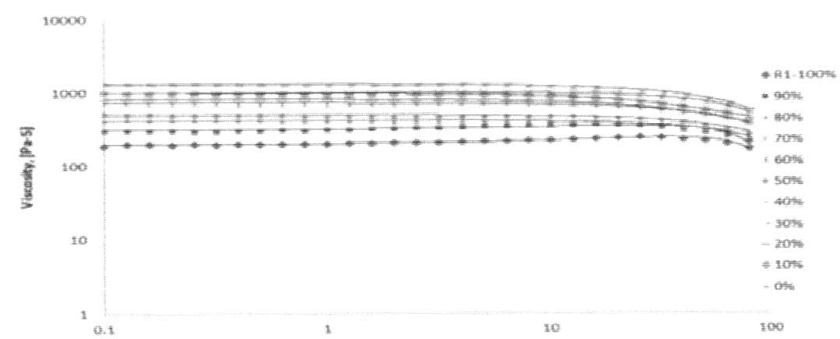

(b)

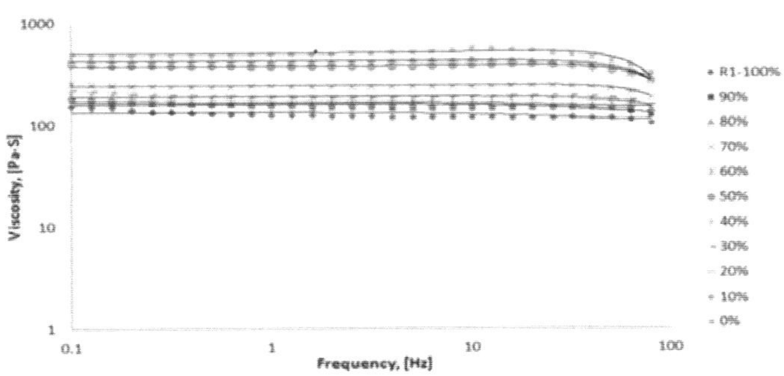

Figure 1. Complex viscosity at a temperature of 255oC: (a) without additives and pigments, and (b)with additives and pigments [36]

The change in viscosity at the higher end of the shear rate range occurred due to shear thinning – the slippage of chains over each other and molecular alignment. The increase in temperature led to a decrease in viscosity because of the mobility of polymer molecular chains increases with temperature.

Table 2. Effect of processing parameters on color in terms of tristimulus values.

Parameters	*°C*	*kg/hr*	*rpm*	*dE**
Temperature (°C)	230	25	750	1.096
	255	25	750	0.346
	280	25	750	0.303
Feed Rate (kg/hr)	255	20	750	0.44
	255	25	750	0.346
	255	30	750	0.320
Screw Speed (rpm)	255	25	700	0.633
	255	25	750	0.347
	255	25	800	0.623

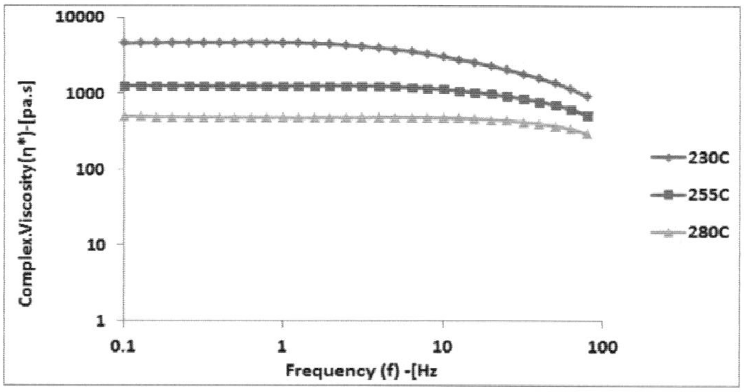

Figure 2. Complex viscosity for R1 30% blends at 255°C

According to **Figure 2,** the viscosity value is less at 255°C than unity at temperatures of 230 °C, indicative of lower viscosity values and greater wettability. Initially, the composites were rheologically characterized using an Ares-G2 rheometer, in three different dynamic modes at 230, 255, and 280°C. **Figure 2** shows the complex viscosity behaviour vs. frequency, measured at a frequency of 10Hz, a. Complex viscosity exhibited mostly constant behaviour in the range from 0.1 up to a strain magnitude of 100%. As expected, complex viscosity decreased as the temperature increased. The relationship between viscosity and strain was constant at lower strains and all temperatures but started to show shear thinning behaviour at higher strains. Also, as the temperature increased, shear-thinning began at the higher strains [39;40]

3.2 Rheological behaviour of PC blends (Micro-TSE) – ML

The rheological properties of polycarbonate blends were characterized without (WOA) and with pigments and additives (WA) while changing the blend proportions in discrete equal intervals. The PC blends were rheologically characterized at 255°C, as shown in **Figure 1,** using the rotational rheometer. The blending of two polycarbonate resins (R1, R2) was studied by modifying the formulations of the pigments and additives in controlled steps; This resulted in eleven batches named ML. A decrease in viscosity due to the presence of additives was observed. An increase in temperature at 255°C decreases the viscosity and improved the surface tension of polymer blends, thereby increasing the wetting properties of pigments. The increase in R2 in the mixture of the two-polycarbonate formulations significantly decreased the melt flow index; the melt flow index for R1 was 28.22 gm/10min, and that of R2 was 8.10 gm/10min. [39]

3.3 Rheological behaviour of compounded plastic (Coperion TSE) – SB

Grades were prepared using SB extruder and injection molding. Output samples were measured by oscillatory rheometry at three levels of temperature, 230,255,280°C. The temperature was found to affect the color difference. It was also determined that SB's viscosity was different from that of ML for the same-compounded plastic.

Lower viscosity improves pigment wetting, lowers color difference and ultimately results in higher dispersion. Thus, the higher pigment dispersion occurred at the point of 255°C or greater. Compounding plastic at 255°C showed that the material exhibited a Newtonian plateau at low frequencies. At a high frequency or shear rate, the viscosity curve exhibited a much more significant decrease and exhibited non-Newtonian behaviour at or above 255°C. Viscosity at 255°C was quite adequate for wetting and deagglomeration of particles. Therefore, 255°C may be regarded as a rheological threshold composition and a good sign for higher colorant dispersion and lower color difference (dE*). In general, as frequency increased, the storage and loss modulus increased, and viscosity decreased. Thus, the reduced viscosity seems to exhibit a correlation with greater pigment dispersion leading to a reduced color deviation. As presented in **Table 2** and **Figure 2**. [36;40]. The blending and mixing process is paramount to the screw design. Therefore, the processing screw design of SB significantly results in a higher viscosity than ML. Compounded plastic's experimental characterization will enable us to perform the design of the proper experiments with consistent parameters. [36-40]

3.4 Effect of temperature on viscosity and color values

Table 2 illustrates the effect of temperature on the color difference (dE*) at 230, 255, and 280°C, for the samples produced at 750 rpm and a feed rate of 25 kg/hr. It indicates a reduced value of color difference as temperature increased. The best color difference value (dE*) occurred at higher temperatures 255 and 280°C, with dE*=0.3.

3.5 Evaluation of pigment dispersion

From analyzing pigment dispersion using the particle size analyzer (PSA) in the wet test, the raw pigment size was broken to lower size values, measured down to approximately 0.1-0.2 µm. Scanning electron microscopy (SEM) produced images to evaluate the pigment dispersion of raw pigment materials. The average pigment size measured approximately 0.1-0.2 µm. Lastly, the thin slices' pigment size results from the injection-moulded chips, imaged with SEM, recorded average particle sizes of 0.2 µm. The obtained measurement test typically agreed with the optimized measurement of the optimal minimum color difference. Furthermore, it could be used as a powerful tool to optimize compounding materials for the color matching process.

3.6 Viscosity and particle size distribution

Further investigation and correlation between viscosity and dispersion were carried out for different screw extruders at a temperature of 255°C. **Figure 3** shows the viscosity (a) and particle size distribution (PSD (b)) measurements for the blends processed in Minilab (ML) and (SB). The blend processed in ML shows a lower viscosity than the one processed in SB. For the ML blend, a larger number of particles (69%) had a small particle size, compared to 50% for SB. This could be attributed to the lower viscosity of the melt, which caused greater wetting, leading to the molecular bonds to break to produce a more significant number of particles. The presence of additives, blending, extrusion and injection molding, high pressure, shear rate and shear heat could also cause the particles to break into smaller sizes. The compounds produced with SB showed a higher viscosity, which could be due to the smaller particle size distributed in the resin matrix. The most favourable result in terms of color was achieved when the blend exhibited higher viscous behaviour. Smaller particle sizes present a higher total surface area and thus give rise to a higher viscosity; therefore, a higher shear heating would not affect the color mismatch and degradation.

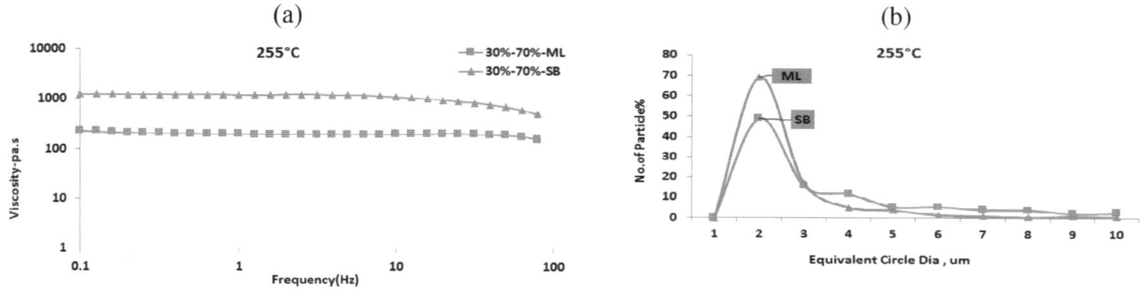

Figure 3. Viscosity (a) and particle size distribution (b) for the blends processed in two different extruders (ML and SB) at a temperature of 255°C

3.7 Morphological dispersion analysis

Figure 4 illustrates the compound grade's SEM micrograph for temperature processing parameters, with a magnification range of 500X-3000X. The pigment particles here came from the 30%-R1 sample. Agglomeration of the pigment occurred at lower processing parameters, e.g. 230°C. The total color difference, dE*, increased at lower processing parameters, and higher agglomerations than that of the sample were produced at higher parameters.

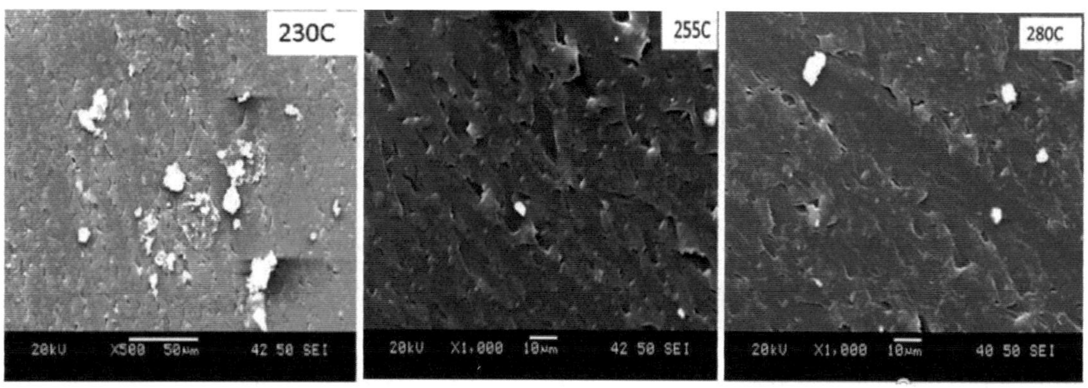

Figure 4. Figure with short caption (caption centred).

At the higher temperature range (255°C and 280°C), the samples were exhibited a lower color difference. According to these results, the sample processed at 255°C showed better-dispersed pigment than that processed at 230°C. Agglomeration can, in principle, take place in zones with low shear. By increasing the temperature to 255°C, deagglomeration occurred in the zones with high shear. By raising the temperature to mid-255°C, the color differences started to decrease. This can be attributed to the relatively low overall shear forces through the entire mixing zones, which separate the composition's pigment particles.

The color difference values (dE*) decreased significantly with the increase in temperature. It showed a consistent higher peak distribution. Equally, large shear forces and frictional heat were possibly generated that may have affected pigment's heat stability and caused damage to the other components of the blend. Consequently, failure in appearance, physical properties, or degradation is possible. Viscosity plays a crucial role in determining pigment wetting, pigment agglomeration, pigment flow and dispersion, and, ultimately, the color shifts. Particle size and color differences are reduced, and a higher peak distribution occurs with increasing the temperature. However, the color difference was reduced at the center level. In general, the color differences were reduced at the center level of the three processing parameters at (i.e. 255°C, 750 rpm, 25 kg/hr), and color output was improved.

4. Conclusion

At the centre level of temperature (255°C). Materials of different formulations processed with different screw configurations were attributes for several applications: It was found that at the centre level of temperature (255°C), smaller sized particles with better dispersion were observed. Scattering efficiency increased rapidly as the average particle size decreased to approximately 0.2 μm, and the optimal number of particles increased. It was also observed that agglomeration occurred in zones of high pigment size. The particle size analyzer (PSA) and imaged with SEM were used in the test to study pigment dispersion. It was found that raw pigment sizes were approximately 0.1-0.2 μm (particle diameter). Thin slices of chips were also imaged with SEM. The obtained measurements were exhibited the minimum color difference. A correlation between viscosity and dispersion was supported for the two screw extruders at 255°C. The blend exhibited higher viscous behaviour in SB than ML. Smaller particle sizes present a higher total surface area and thus give rise to a higher viscosity; therefore, a higher shear heating would not affect the color and degradation. The particle size distribution (PSD) measurements for the blends processed in Minilab (ML), Coperion extruder(SB), and the (SEM) micrograph of the compound grade were used to test the pigment dispersion, the pigment size was decreased to approximately 0.2 μm, deagglomeration occurred in the zones with high shear, and exhibited the minimum color difference *(dE*)=0.3)*. The color values (dE*) decreased significantly by raising the temperature to mid-255°C and color output was improved.

5. References

[1] Syang-Peng, Rwei. Polymer Engineering and Science, 2001
[2] Wong, A and Lam, Y, J Polym Res 15:11–19, 2008
[3] Mishra, A, Kumar. Scientific Journal & Industrial Research 67 (2008).
[4] Dealy, J, and Wissburn, KF. In Melt Rheology and its Role in Plastics Processing: Van Nostrand New York: Kluwer Academic Publishers, 1990.
[5] Abdel-Goad, M, and Pötschke, P."Rheological characterization of melt-processed polycarbonate multiwalled carbon nanotube., pp: 2-6. 2005.
[6] Mange Sana, N, Chi kuku, RS, AN." Journal of the South African Institute of Mining & Metallurgy / Saimm, 108, no. 4 (2008): 237-243.
[7] Zaeh, M and Hess, R. "Impact of stabilizers on the color of pigmented polymer articles Geneva." Switzerland, Additives Geneva, 2002.
[8] Sanchez, P. M. Remiro, Nazabal, J J. Appl. Polym. Sci., 50, 995-1005 (1993)
[9] Liang, R. F. and. Gupta, R.K, 2000, Society of Plastic Engineering ANTEC, 2903 2907.
[10] M. Khan, "Rheological and mechanical properties of ABS/PC blends," Korea-Australia Rheology Journal, vol. 17, no. 1, pp. 1-7, 2005.
[11] Lee S, Mather P, Pearson DMMK Journal of applied polymer science 59, 243-250, 1996
[12] Fatoni, Rois. Product Design of Wheat Straw Polypropylene Composite. Waterloo: University of Waterloo, 2012
[13] D. Meade, Introduction to colorant selection and Application Technology, in Coloring of Plastics Fundamentals, 2nd ed., R. Charvat, Ed., 2004.
[14] B. Mulholland, "Effect of Additives on the Color & Appearance of Plastics," ANTEC, Ticona Engineering Polymers, 2007.
[15] J. Birmingham, "Volatility of Titanium Dioxide Pigments in Polyethylene Film Extrusions," ANTEC, p. 3290–3294, 1995.
[16] E. Jaffe, C. Campell, S. Hendi and F. Babler, "Rheologically Effective Organic Pigments," J.Coating Technol, p. 47–54, 1994.
[17] D. Williams and M. Bevis, "The Effects of Recycled Plastics and Compound Additives on the Properties of Injection Polypropylene Sci, p. 2834–2842, 1980.
[18] A. Mishra and Kumar, Scientific Journal & Industrial Research, vol. 67, 2008.
[19] D. J and K. Wissburn, In Melt Rheology and its Role in Plastics Processing: Theory and Applications Van Nostrand Reinhold, New York: Kluwer, 1990.
[20] K. H, H. Lee and J. Wook, "Rheological properties of branched polycarbonate prepared by," Korea-Australia Rheology Journal, vol. 19, no. 1, pp. 1-5, 2007.

[21] P. R. Hornsby, "Rheology, Compounding and Processing of Filled Thermoplastics," Mineral Fillers in Thermoplastics I, pp. 155-217, 1999.
[22] A.-G. M and P. Pötschke, "Rheological characterization of melt-processed polycarbonate multiwalled carbon nanotube composites," Fluid Mech, pp. 2-6, 2005.
[23] C. K and W. Cook, "Structure-Property Relationships of Blends of Polycarbonate," Polymer engineering and science, vol. 43, no. 11, pp. 1727-1739, 2003.
[24] B. V and P. Kitchloo, "Rheological characterization of polycarbonate resins and its applications," Society of Plastics Engineers, pp. 103-106, 2011.
[25] H. S, K. Tapan and S. Chattopadhyay, "Dynamic and capillary rheology of LDPE-EVA–based thermoplastic elastomer, vol. 31, no. 3, p. 377–391, 2010.
[26] S. E, A. Hossein and Behravesh, "Rheological Investigation of Wood-Polypropylene Composites " Journal of Polymers and the Environment, pp. 998-1006, 2012.
[27] A. S, W. Leelapornpisit, A. Derdouri and P. Carreau, "Morphology and Properties of Microinjected Polycarbonate/MWCNT Nanocomposites," in 66th Annual Technical Conference of the Society of Plastics Engineers, 2008.
[28] N. Mangesana, R. Chikuku and A. Mainza, "The effect of particle sizes and solids concentration on the rheology of silica sand-based suspensions Journal of the South African Institute of Mining & Metallurgy / SAIMM, vol. 108, no. 4, pp. 237-243, 2008
[29] M. Zaeh and R. Hess, "Impact of stabilizers on the color of pigmented polymer articles Geneva," Switzerland, Clariant Business Unit Additives - Geneva, 2002
[30] J. E. Pickett. In-Service Life Prediction: Challenging the Status Quo; J. W. Martin; R. A. Ryntz; R. A. Dickie, Eds., 2005; pp 93-106.
[31] A. Factor; W. V. Ligon; R. J. May. Macromolecules 1987, 20, 2461-2468.
[32] P. Hrdlovic. Polymer News, 2004, 29, 187 -193.
[33] J.S. Humphrey Jr; A. R. Shultz; D. B. G. Jaquiss. Macromolecules, 1973, 6, 305-314.
[34] H. Zweifel. Stabilization of Polymeric Materials; Springer, 1998.
[35] ASTM Standard D2244, 1993 (2000), “Standard Test Method for Calculation of Color Differences,” ASTM International, West Conshohocken, PA
[36] J. Alsadi., Color Mismatch in Compounding of Plastics: Processing Issues and Rheological Effects (Doctor of Philosophy in Mechanical Engineering. The Faculty of Engineering and Applied Science University of Ontario Institute of Technology). Ontario, Canada (2015).
[37] J. Alsadi, U. Saeed, S. Ahmad, G. Rizvi and D. Ross., Processing issues of color mismatch: Rheological characterization of polycarbonate blends, Polymer Engineering & Science, Volume 55, Issue 9, September 2015, Pages: 1994–2001, Version of Record online: 8 DEC 2014.
[38] J. Alsadi, Investigation of the effects of Formulation, process parameters, Dispersions, and Rheology on using combined Modelling and experimental Simulations, published in Materials Today: Proceedings and published on April 2, 2019, ELSEVIER. Paper May 22, 2019. Volume 13, Part 3, 2019, Pages 530-540
[39] J, Al Sadi, Systematic review: Impact of Processing Parameters on Dispersion of Polycarbonate Composites, and Pigment Characterized by Different techniques, submitted on April 3, 2020, to Materials today: Proceedings Elsevier, Published on June 10, 2020. Volume 27,issue 4, pp -3254-3264 ,DOI information: https://doi.org/10.1016/j.matpr.2020.05.027
[40] J. Alsadi, “Materials Processing Issues and Experimental Study for Rheology and Dispersion of Polycarbonate Grades: Twin Screws design.” the manuscript has been accepted to the Second International Conference on Industrial, Systems & Manufacturing Engineering (ISME’19)” arranged by, Jordan Engineers Association (JEA) and German Jordanian University (GJU), accepted on October 17, 2019, Presented on November 11-13, 2019

Acknowledgments
The author would like to extend his gratitude for the support provided by Engineer Nour Al Sadi at Ontario technical university, Canada for excellent assistance.

Voice Recognition to Control and Monitor Robotic Arm on IoT by Using Smartphone

Ngoc-Hoang Tran

University of Technology and Education – The University of Danang, Danang, Vietnam

tnhoang@ute.udn.vn

Abstract. Robotic control approaches are extensively studied and developed in recent years, especially in industrial manufacturing. They are explored in various way from using image-processing to voice-controlled application. Robot communication techniques are required to develop by three dimensions of production: better quality connection; safer and cheaper. In this context, this paper proposes a protocol of control and monitor Robotics by using a developed technology of the 4.0 industries: Internet of Thing (IoT). Control system by voice recognition using Google Assistant on Smartphone and IoT communication by using Adafruit IO with IFTTT platform as data transfer intermediaries. Our approach applies to control an own developed robotic arm with Arduino Uno as the controller. Our paper show also the voice controlling commands are successfully transmitted via IoT technology and the desired operations are performed successfully and also analyse its advantages and disadvantages and the perspectives.

1. Introduction

Today, robots take an important role in the science and technology development. Indeed, robots have been widely applied in military, industrial and agricultural techniques to replace people's role in dangerous environments. Many technologies are developed for controlling robot in recent years. Among them, Voice – recognition technique is proved as an intelligent method that was increasing widely role in this context such as Google Assistant, Alexa, Cortana [1].

Following, a large number of smart/intelligent robot was designed by using wireless communication. They express the human-robot interaction (HRI) term with more productivity, less energy resource waste and more convenience than old control methods by physical communication (keyboard, computer, LAN, etc.). For example, the researches [2,3] uses Bluetooth to communicate/control to robots from the Blue-tooth transmitter to command robot to perform operations. These studies use mainly circuit systems with old sensors such as RSC364 (used in radio transmission) to analyse the voice frequency to command. This approach faces disadvantages such as controlling is possible in low distance depends system, high cost for circuit construction and its quality depends on the recording quality of the microphone.

Furthermore, [4,5] propose a new Bluetooth transmitter approach via smartphone. The control design in these works was simpler than the old type. By using WIFI, 3G, 4G, or LAN network, communication with the Robot is significantly improved in distance terms, more flexibility and more convenience [6]. However, they are still limited at storage capacity, direct / indirect connection and poor system security. In our part, we propose an advanced approach to control robot by using IoT communication. Voice recognition command through Google Assistant, transmit/feedback IoT by intermediates servers. By this way, this approach has many important ad-vantages that are more secure and safely; controlling robot is not depend on physical distance; easy to access for monitoring by historical data on server. This article is structured as following: the next section proposes the control/monitor robot process. This section presents also the used techniques in this work. Based on this, we show our results that tested on an own developed robotic arm. The conclusion and the perspective will close the article.

2. Proposed Control and Monitor Process

This process is represented by three main parts: Human - Voice command on Smartphone environment; IoT platform for transmit order and receive feedback from sensor system and robotic arm controlled by Arduino Uno, as described as figure 1.

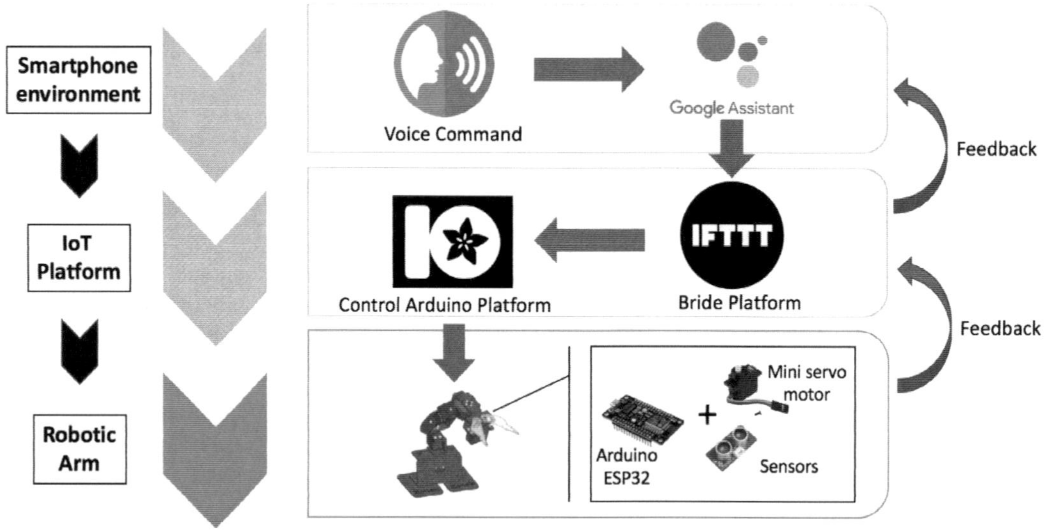

Figure 1. Proposed process of controlling and monitoring robotic arm by voice command on IoT

2.1. Human – Voice command on Smartphone environment

To convert voice to text or computer commands, it must be performed by a complex process with many steps. Basically, an Analog-to-Digital Converter (ADC) converts these analog (analog) waves (human voice) into data that system can understand, then filtering (separate) noise into different frequency bands to eliminate, see more details in [7]. Fortunately, all of these complex processes are supported by Google Assistant. Google Assistant system will have to receive sound from the human speaking on the phone, the phone will record the monologue through google, the data will be processed as text characters.

2.2. Establishment of IoT platforms

The communication module used in this work was the advanced development protocol, which is a virtual interface platform that communicates with the ESP32 chipset: Adafruit IO. This is an IoT platform built around the Message Queue Telemetry Transport (MQTT) Protocol [8], that has extensive literature focused on IoT solutions, all developed by the Adafruit Industries.

In other side, If-This-Then-That (IFTTT) is a widely used platform with over 11 million users [9] which is considered as a useful bridge between Google Assistant and Adafruit IO platform. IFTTT platform is designed as a trigger-Action platform which support stitching together various online services APIs such that end-users may write simple conditional programs. These simple programs often take the form "IF triggering condition, THEN take a specific action". We designed the synchronization operation of these two servers. When a voice command is recorded on Google Assistant, it is encoded and transmits this encrypted signal into text form from Google's API through ITFFF Server to Adafruit IO and sent directly to Arduino ESP 32.

2.3. Robotic Arm controlled by Arduino

Arduino is an open-source platform based on hardware (8-bit AVR Atmel Processor or 32-bit ARM Atmel) and software represented correspond as a simple I/O board (equipped with a USB interface, 6 analog input pins, 14 digital I / O pins allow to expand with various boards) and a Processing/Wiring language developed environment (called Arduino Integrated Development Environment, also Arduino IDE). This Arduino IDE is published as open source tools, available for extension by experienced programmers. The language can be expanded through C++ libraries to user easily the AVR C programming language on which it's based. In our paper, the robotic arm controlled by Arduino is designed with 3 Degree of Freedom (DOF) as described in figure 2 and table 1.

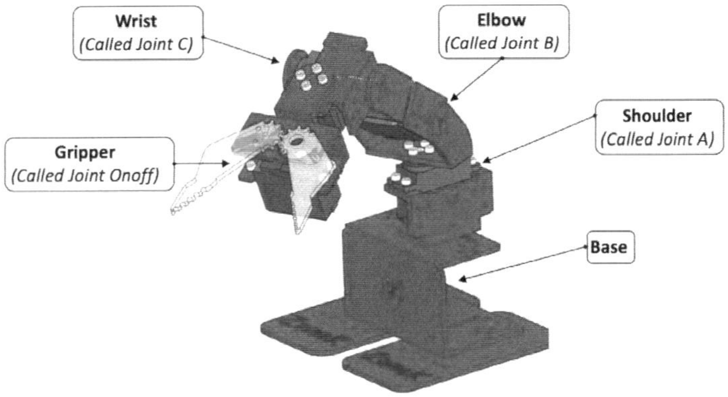

Figure 2. 3DOF Robotic Arm design model.

Table 1. Robotic Arm Specifications

Names	Descriptions
Overall size	250 x 60 (cm)
Material	Metal
Processor	Arduino ESP32 (5V – 90mA)
Motor	Mini Servo (with feedback signal)
Sensor	IR Infrared Proximity (5V – 15mA)

3. Results and discussions

Following to proposed process, following steps describe our implementation and its results:

1) Create an applet editor (on ITFFF platform), where we choose triggers ("If This") and the subsequent actions ("Then That"). We choose in the next "Google Assistant" as the service for our trigger, then set up its settings such as: the Google Assistant's response, language, etc. In our case, the phrase "Ok I'm in processing" to response command order. It implies that when the ESP processor is received the request and is performing actions. The final part of this applet is the Action, we choose "Adafruit", meaning to "Send data to Adafruit IO". Till this applet is completed, the converted voice-command to text from Google Assistant will be sent to Adafruit IO platform in which transmit to microcontroller. Figure 3 show us an example of voice - command by Google Assistant to ESP processor and its response.

2) Create a new feed of Adafruit IO: In which we create threes "postData" variables correspond to threes joints of Robotic Arm and "onoff" variable to represent feedback values from ESP processor while it is performing.

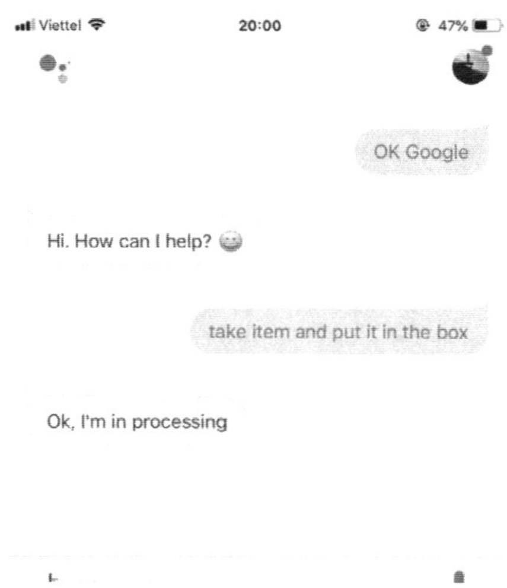

Figure 3. Voice – Command to Robotic Arm by Google Assistant.

In addition, ESP32 processor connect to Adafruit IO by MQTT Protocol. It's performed by to install the Adafruit MQTT Client library, which can be found under the Arduino Library Manager [10] by including at the start of our program:

```
#include <ESP32-WiFi.h>
#include "Adafruit_MQTT.h"<br>#include "Adafruit_MQTT_Client.h"
```

Once the connection is completed, ESP processor will receive the voice-command from Google Assistant through above step 1 and Robotic Arm perform actions corresponds. Sensor from mini-servo feedbacks to Adafruit Server to monitor Robotic Arm's activities. Figure 4 prove us controlling and monitoring Robotic Arm's performing process.

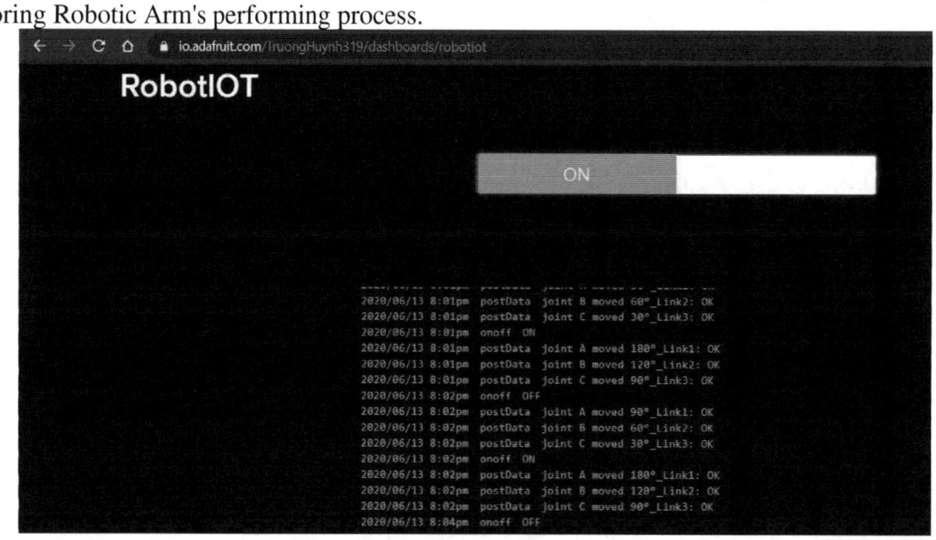

Figure 4. Interface of monitoring of Robotic Arm's performance on IoT server.

In figure 4, information feedback from joint A, B, C and onoff correspond to the components of Robotic Arm described in figure 2. They are updated automatically and continuously on Adafruit IO server to monitor in real-time.

3) Finally, these feedbacks are synthesized in our Adafruit server and send a notification on our smartphone when Robotic Arm has finished its activities, as described in figure 5. The results show us that the control and monitoring of the robot's execution are successful. The feedback from the robot's sensors (~5-10ms/cycle) almost no-delay.

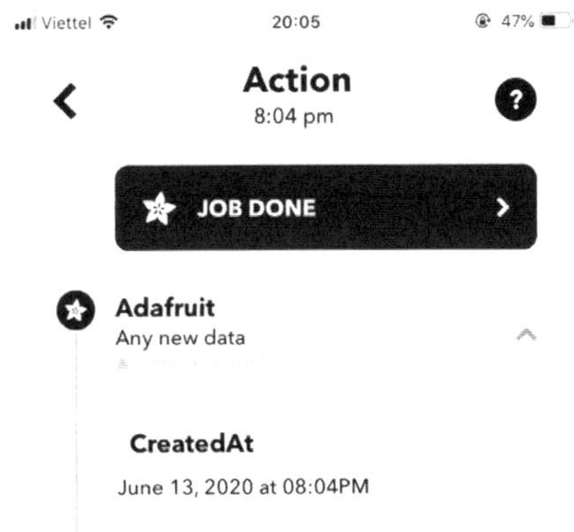

Figure 5. Successful notification of Robotic Arm on Smartphone.

However, this work still has some limitations: (i) Voice - recognition is depended on Google Assistant in which it has not specifically designed for each individual language. Moreover, the extension with the effective filters to improve recognition quality is impracticable. (ii) Using two common Adadruit and IFTTT platforms will be difficult to avoid security risks, and its connections are unstable.

4. Conclusions and future works

Our contribution in this article to provide an advanced approach of controlling and monitoring robot's operations. Voice commands are successfully transmitted from Google Assistant on Smartphone through IoT cloud to Arduino processor. User will get notification while desired activities has done successful. This research reduces human activities in dangerous places or situations: such as radioactive mineral exploration, geological measurement. In perspective, an extension monitoring protocol with camera system is necessary to developer on integrating with our methodology. In addition, robot positioning with build self-propelled algorithm make improve supervision system.

5. References

[1] López, Gustavo, Luis Quesada, and Luis A. Guerrero. 2017. Alexa vs. Siri vs. Cortana vs. Google Assistant: a comparison of speech-based natural user interfaces. *International Conference on Applied Human Factors and Ergonomics. Springer, Cham.*

[2] Abdullahi, Zakariyya Hassan, et al. 2015. Mobile robot voice recognition in control movements. *International Journal of Computer Science and Electronics Engineering (IJCSEE)* 3.1: 11-16.

[3] Fezari, Mohamed, and Mounir Bousbia-Salah. 2006. A voice command system for autonomous robots' guidance. *9th IEEE International Workshop on Advanced Motion Control, IEEE.*

[4] Kahar, Saliyah, et al. 2011. Data transferring technique for mobile robot controller via mobile technology. *International Conference on Pattern Analysis and Intelligence Robotics*. Vol. 1. IEEE.

[5] Ersahin, Gonca, and Herman Sedef. 2015. Wireless mobile robot control with tablet computer. *Procedia-Social and Behavioral Sciences* 195: 2874-2882.

[6] Hassan, Mohammad Ashri Abu. 2012. A review of wireless technology usage for mobile robot controller. *Proceeding of the International Conference on System Engineering and Modelling (ICSEM).*

[7] Andics, Attila, et al. 2010. Neural mechanisms for voice recognition. *Neuroimage* 52.4: 1528-1540.

[8] Kodali, R. K., and Kopulwar S. M. 2016. A low-cost implementation of MQTT using ESP8266. *2nd International Conference on Contemporary Computing and Informatics (IC3I)*. IEEE.

[9] IFTTT, 2011. https://platform.ifttt.com.

[10] Maguluri, Lakshmana Phaneendra, et al. 2018. Efficient Smart Emergency Response System for Fire Hazards using IoT. *International Journal of Advanced Computer Science and Applications (IJACSA)* 9.1: 314-320.

Numerical Simulation and Experimental Research on Engine Room Heat Dissipation Performance of a Hybrid Car

WANG Jianpeng[1], **YAN Tingsong**[2], **ZHANG Zhijun**[2]

(1.BYD Company Limited, Product Planning and New Auto Technologies Research Institute, Shenzhen, 518118, China;
2.School of Mechanical Engineering and Automation, Northeastern University Shenyang,110819, China)

987179244@qq.com

Abstract. Compared with the fuel cars, the hybrid cars has higher thermal load of engine room and has higher requirements for heat dissipation performance of the cooling module. Under idling condition, a hybrid car has an actual problem that higher temperature of engine room and inlet air temperature of cooling module. In this paper, a mathematic model of car thermal management is established, numerical simulations on flow and temperature fields of car were performed by using STAR-CCM+ software, the feasibility of the simulation results was verified by the thermal balance test of the cars. The results show that the reflow of hot air and the inlet air temperature of the cooling module are reduced by sealing the cooling module, the front end module and optimizing the structure of deflector, thus, the heat dissipation efficiency of the cooling module is improved.

1. Introduction

Compared to fuel cars, the engine room of hybrid cars need to be equipped with parts such as engine, generator, gearbox, motor and controller at the same time, the heat dissipation condition is worse. Therefore, research on engine room heat dissipation performance of hybrid cars has become the focus for domestic and foreign scholars. CFD approach is applied to all aspects of car design, it can significantly shorten the product development cycle and save R&D costs [1-3]. Kim J M [4] performed a numerical simulation on the shape of the grille opening, the influence of structural parameters such as the vertical height, horizontal width and position of the grille on the air intake flow is analyzed, which provides reference for the improvement of heat dissipation performance of car engine room. Ma W S [5] tested the heat dissipation efficiency of a hybrid car radiatos, the influence of different air intake speed and temperature under multiple working conditions on the heat dissipation efficiency of the cooling system is regularized. Vivek Kumar [6] analyzed the flow and temperature field in the engine room and verified the simulation results through test, it provided a theoretical basis for the optimization of the car cooling system. Shi Pengfei [7] used CFD approach to simulate the flow field of engine room and found that the air intake flow of the cooling module could be effectively increased by adding deflectors. Patidar [8] found that the reflow of hot air was an important reason for low heat dissipation efficiency of cooling module and overheating of engine room. Li Zhelong [9] restrained the reflow phenomenon of hot air and improved the cooling efficiency of the condenser by configuring the condenser bossing and adjusting the opening of the grille.

Most of the research is mainly to simulate the air flow and temperature field of traditional fuel cars, the research on the heat dissipation characteristic of the hybrid car engine room is rare. This paper aiming at the actual problems of high inlet air temperature of the cooling module and high temperature of engine room under idling condition, CFD approach is used to analyze the causes of problems and propose improvements. Moreover, the correctness of the simulation results are verified through the thermal balance test so that making the thermal management performance of the engine room meets the requirements.

2. Establishment of numerical simulation model

2.1 Numerical model

Under idling condition, mach number is less than 0.3, the air density around the car can be approximated as a constant. The air flow field in the engine room is complicated, the standard $k-\varepsilon$ model is selected. The basic equations are as follow.

Continuity equation:

$$\frac{\partial}{\partial x_j}\left(\rho u_j\right)=0 \tag{1}$$

Momentum equation:

$$\frac{\partial}{\partial x_j}\left(\rho u_j u_i-\tau_{ij}\right)=-\frac{\partial p}{\partial x_j} \tag{2}$$

Energy equation:

$$\frac{\partial}{\partial x_j}\left(\rho u_j h-k\frac{\partial T}{\partial x_j}\right)=-u_j\frac{\partial p}{\partial x_j}+\tau_{ij}\frac{\partial u_i}{\partial x_{ij}} \tag{3}$$

Where ρ is the density, T is the temperature, u_i is the velocity component of i direction on the coordinate axis, p is the pressure, τ is the viscous stress, k is the heat transfer coefficient, h is the fluid enthalpy.

2.2 Simulation model

The size of computational domain is consistent with the thermal balance laboratory. The whole car has 35 million grids. The computational domain is shown in Figure 1. The boundary conditions of the simulation model are set according to the idling condition of thermal balance test. The computational domain uses velocity inlet and pressure outlet. The surfaces of exhaust system, engine, gearbox and other use temperature wall boundary condition. The heat protection shields are set as solid area. Each radiator is set as porous medium area. MRF model is applied to the fan. Radiation model uses S2S model.

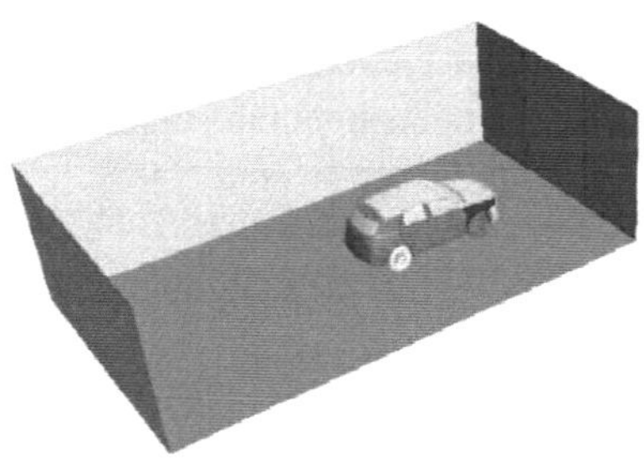

Figure 1. Computational domain

3. Analysis of simulation result

Under idling condition, the airflow tends to form partial reflow around the cooling module. The cooling airflow passes through the grille, motor radiator, condenser, low temperature radiator, high temperature radiator and fan and then blows to the surfaces of engine, motor, controller, gearbox and exhaust pipes. The airflow cools the radiators firstly and then carries out convective heat transfer with the engine room parts to achieve the goal of the heat dissipation.

3.1 Analysis of original model

(1) The maximum spacing between the upper, lower, left and right faces of the cooling module and the front end module is 20mm, 24mm, 23mm and 22mm respectively, as shown in Figure 2. (2) There is a spacing between the radiator groups. (3) The deflector is divided into four pieces on the left and right sides, which is fixed on the radiator water chamber and the front end module. The front of deflector is connected with anti-collision beam and front bumper, there is gap between the deflector and water chamber, anti-collision beam and front bumper. The deflector is shown in Figure 3.

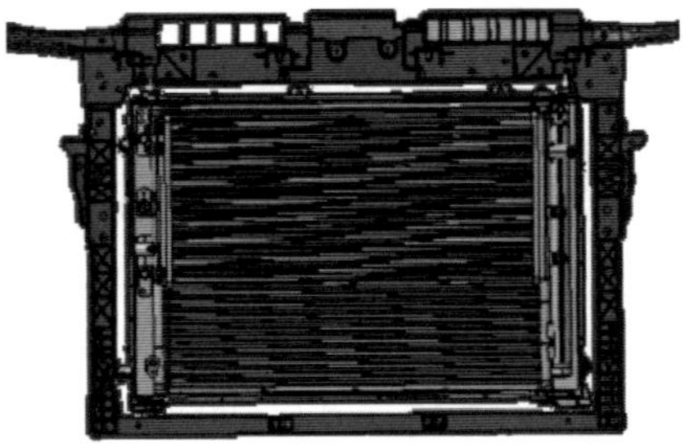

Figure 2. Cooling module and front end module

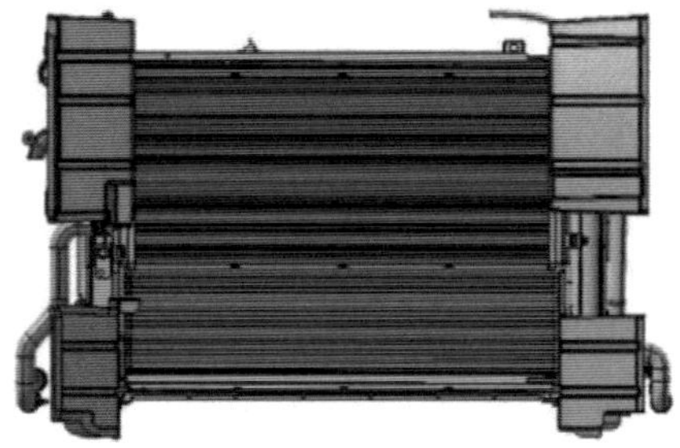

Figure 3. Deflector

Figure 4 shows the air reflow in the Y=0 mm section of engine room. After airflow passes through the radiators and fan, part of airflow passes through the gap between the top and bottom of cooling module and the front end module and flows back to the upper front and lower front of cooling module. Figure 5 shows the air reflow in the Z=260 mm section. Part of the airflow passes through the gap between the cooling module and the front end module, the front end module and front bumper, in the end, it flows back to the front of cooling module.

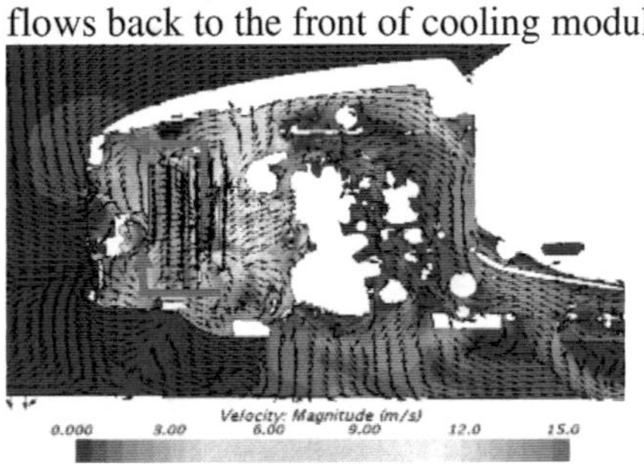

Figure 4. Y=0 mm

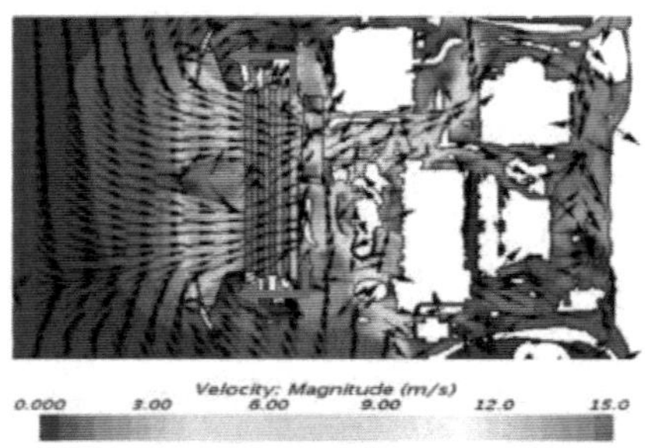

Figure 5. Z=260 mm

Monitor the air intake flow and inlet air temperature of the grille and radiators separately, analyze the influence of hot air reflow on cooling effect. The monitoring air intake flow position of grille is 5 mm away from the grille, the monitoring air intake flow position of radiators is 2 mm away from the surface of radiators.

Table 1 shows the hot air reflow rate around the cooling module is 38.47%, which seriously affects the Wind absorption efficiency of the fan. Table 2 shows the inlet air temperature of the grille is 45.17°C, the motor radiator is 50.85°C, which are higher than the corporate standard. After the cooling air which is heated by the radiators enters the engine room, part will reflow to the radiators for reheating again, which obviously worsens the inlet air temperature of cooling module.

Table 1. The air intake flow of grille and radiators(kg/s).

Grille	Motor radiator	Condenser	Low temperature	High temperature

			radiator	radiator
0.443	0.238	0.550	0.596	0.720

Table 2. The inlet air temperature of grille and radiators (°C).

Grille	Motor radiator	Condenser	Low temperature radiator	High temperature radiator
45.17	50.85	58.28	69.40	71.25

3.2 Thermal balance test

In order to verify the correctness of the simulation results, it is necessary to test the original car according to the thermal balance test scheme. The collection is stopped after 5 minutes of thermal balance, as shown in Figure 6.

Figure 6. Thermal balance test

Some test data and simulation results are shown in Table 3. The error between the simulation results and the experimental data is less than 15%. It can be considered that the CFD simulation results are feasible.

Table 3. Part of test data and simulation results (°C).

Position	Grille	Motor radiator	Condenser	Rear bumper	Rear lifting lug
Simulation value	45.17	50.85	58.28	46.14	93.13
Test value	49	58	65	52.6	105.4
Error	7.82%	12.33%	10.34%	12.28%	11.64%

4. Optimization of heat flow field in engine room

4.1 Optimization analysis

Under idling condition, not only the hot air reflow around the cooling module be considered, but also the air leakage between radiator groups. (1) Seal the gap between the radiator groups with sponge sealing strips. (2) Seal the gap between the cooling module and the front end module with sponge sealing strips, as shown in Figure 7. (3) Figure 8 shows the new deflector is based on the original deflector by adding two pieces of upper and lower ones. In addition, the gap between the deflector and the radiator water chamber, the anti-collision beam and the front bumper is filled. Compared with Fei hongqing[10], the optimization scheme proposed in this paper has higher feasibility and lower cost to increase the grille intake area and adjust the relative position of the cooling module.

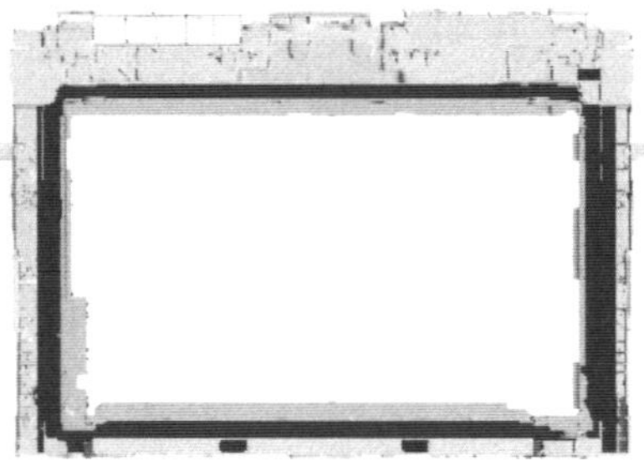

Figure 7. Sealing strip

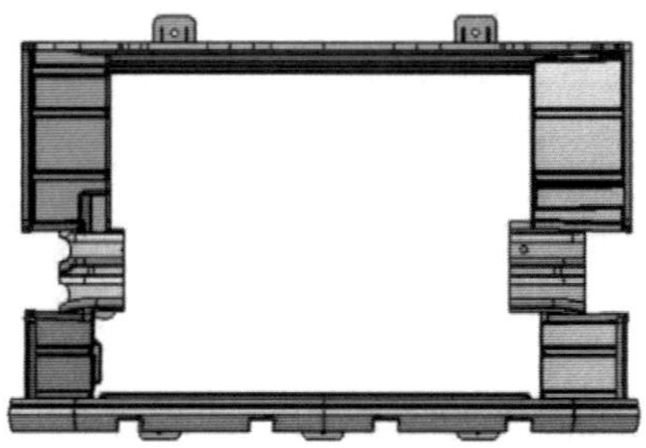

Figure 8. New deflector

Figure 9 shows the flow field distribution in the engine room of the optimization. After the new deflector was replaced and the sponge sealing strips were added, the hot air reflow at the top and bottom basically disappeared. figure 10 shows that after the gaps between the deflector and the radiator water chamber, the anti-collision beam, the front bumper are filled with sponge sealing strips, the reflow of hot air on the left and right sides is obviously suppressed.

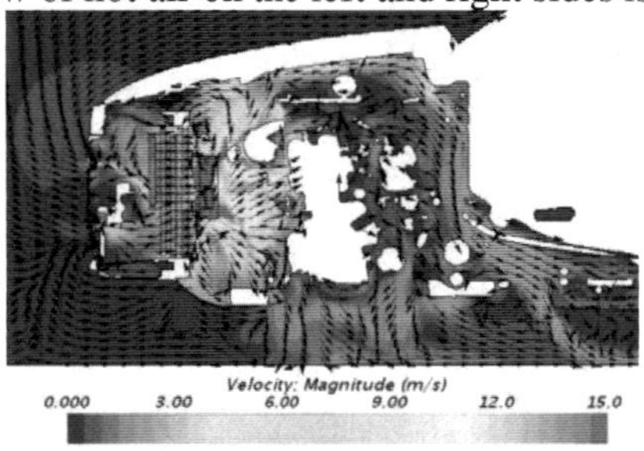

Figure 9. Y=0 mm

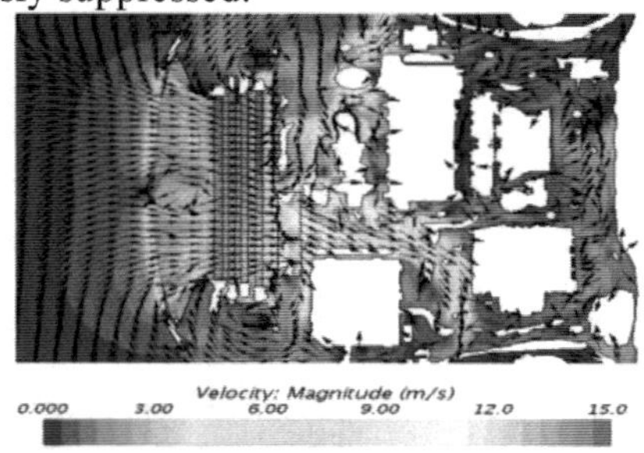

Figure 10. Z=260 mm

The air intake flow of the optimized grille and each radiator is shown in Table 4. The air intake flow of grille, motor radiator, condenser and low temperature radiator has increased obviously and the high temperature radiator is slightly reduced. The hot air reflow rate around cooling module is 11.92% because of the reflow phenomenon was suppressed. Table 5 shows the inlet air temperature of the optimized grille and each radiator has declined in varying degrees;

Table 4. The air intake of the optimized grille and radiators(kg/s).

Grille	Motor radiator	Condenser	Low temperature radiator	High temperature radiator
0.554	0.256	0.629	0.629	0.629

Table 5. The inlet air temperature of the optimized grille and radiators (°C).

Grille	Motor radiator	Condenser	Low temperature radiator	High temperature radiator
45.08	46.44	51.91	61.44	62.05

4.2 Comparsion of results

From the comparison of figures 11 and 12, the average temperature of airflow in the engine room is significantly reduced after optimizing, the inlet air temperature of the motor radiator, condenser, low temperature radiator and high temperature radiator is reduced by 4.41°C, 6.37°C, 7.96°C and 9.2°C respectively. The air intake flow of grille has increased by 25.06%. The hot air reflow rate of the cooling module has been reduced by 26.55%. Reference of 10 shows the inlet air temperature of cooling module drops by 2°C. Therefore, the optimization scheme in this paper has obvious advantages.

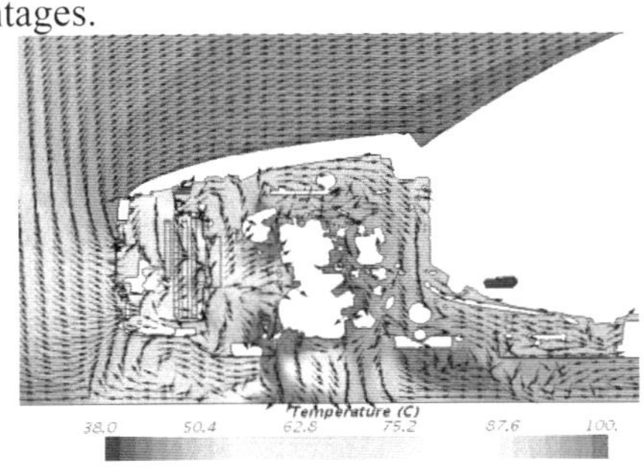

Figure 11. Original car

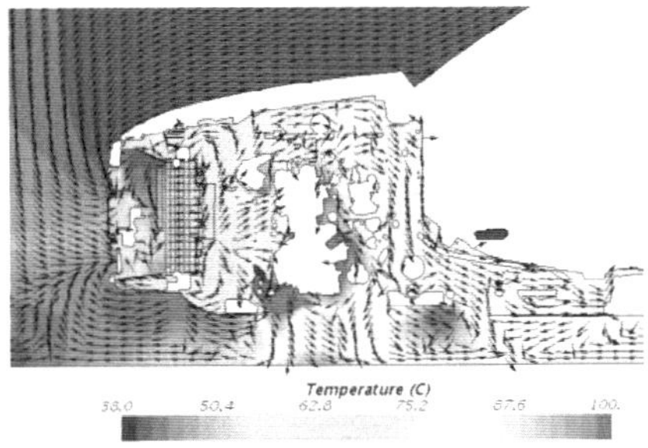

Figure 12. Optimazation

5. Conclusions

In this paper, a combination of numerical simulation and experimental research was used to study the influence of the sealing condition of a hybrid car cooling module on the heat dissipation characteristic of the engine room. Comparing the simulation results of the original status and the optimization with the data of thermal balance test. The conclusions are as followed:

(1) In the original state, the hot air reflow rate around the cooling module is 38.47%. By sealing the gaps of cooling module and front end module and optimizing the deflector structure, the hot air reflow can be effectively suppressed.

(2) After optimizing, the air intake flow of grille has increased by 25.06%, which can effectively improve the heat exchange efficiency of the radiator. The inlet air temperature of each radiator of the cooling module is significantly reduced. The average temperature of the airflow in the engine room drops by about 10°C.

(3) Comparing the thermal balance test data with simulation results, the errors of the simulation are within 15%, the correctness of the simulation results is verified.

6. References

[1] Kepeng Z, Peiyu C, Jun L, et al. Analysis on Thermal Management of a Hybrid Vehicle Engine Compartment[J]. Auto Engineer, 2019.

[2] Llhan Bayraktar. Computational simulation methods for car thermal management[J]. Applied Thermal Engineering,2011,36.

[3] Fortunato F, Damiano F, Matteo L D, et al. Underhood Cooling Simulation for Development of New cars[C]// car Thermal Management Systems Conference & Exposition. 2005.

[4] Kim J M, Kim K M, Ha S J, et al. Grille design for passenger car to improve aerodynamic and cooling performance using CFD technique[J]. International Journal of Automotive Technology, 2016, 17(6):967-976.

[5] Ma W S, Shen W X, Zhang L W. Heat rejection efficiency research of new energy car radiators[J]. Iop Conference, 2018, 324.

[6] Kumar V, Kapoor S, Arora G, et al. A Combined CFD and Flow Network Modeling Approach for car Underhood Air Flow and Thermal Analysis[C]// SAE World Congress & Exhibition. 2009.

[7] Shi P, Yu J,Li F, Liu H 2018 *Auto Eng* **01** 41.

[8] Patidar A, Gupta U, Marathe N. Optimization of Front-End Cooling Module for Commercial car Using CFD Approach[C]// Symposium on International Automotive Technology. 2013.

[9] Li Z, Li T and Li X 2019 *Chin. J. Auto. Eng.* **6** 400.

[10] Fei H, Tian J, Lei S, Chen Y and Yan W 2014 *Internal Combust. Engine Powerplant* **06** 25.

Frequency Damping Backpack Exoskeleton with Load Variable Spring Stiffness

C R Patil[1]*, A Vyas[1], A M Pimpley[1], P S Borkar[1], A Thapa[1], G N Kotwal[2]

[1]Department of Mechanical Engineering, Vishwakarma Institute of Technology, Pune-411037, India
[2]Department of Industrial & Production Engineering, Vishwakarma Institute of Technology, Pune-411037, India

*Corresponding author: chinmay.patil17@vit.edu

Abstract. Heavy backpacks are one of the major reasons for spinal, neck and shoulder related problems .It also affects the human posture in an adverse way. This a major concern among school children, who have to carry heavy textbooks and stationery material. The suspended load backpack is designed to tackle these problems. However the major flaw in these backpacks is the fixed spring constant despite a range of various loads being acted upon it. Furthermore, the current frame is heavy and uses a complex damping technique. In this paper, we make a mechanism to vary the spring stiffness of the backpack according to its payload without having to adjust the damping coefficient. The redesigning of the straps ensures even load distribution during running, jumping, etc. We use a power screw type mechanism coupled with a spring to add preload. We also simulate the model to validate and differentiate between the performances.

1. Introduction

The Floating Backpack eliminates the accelerative forces that cause injuries and reduce mobility. The heavy backpacks kids are made to carry deteriorate their posture. Also, is known to give them chronic problems. Spine is made of 33 bones called vertebrae. Between the 33 bones of the spine i.e. the vertebrae, there exist disks that impose as natural shock absorbers. Improper posture while carrying heavy loads can cause various complications. Due to the heavy weight, they get pulled backwards . To make up for this unbalanced force, they may bend forward at the hips or arch their back. This compensation for improper distribution causes their spine to compress abnormally. This eventually causes the shoulders and the upper back to become curved unnaturally. All of these factors contribute towards discomfort or an injury toward the neck and backside of the. Kyphosis, the slouching of the back, puts additional strain on our muscles. Ignoring physical withdrawals like soreness and discomfort in the back, can lead to persistent health issues [1]. Erroneous posture puts a lot of stress on the muscles which subsequently makes the body work harder to maintain a vertical posture instigating the person to feel exhausted. When the body is in an abnormal position, it incessantly tries to work back to where it's supposed to be. This constant work that the body puts in to correct the posture consumes energy which instigates the feeling of fatigue.

Using a simple suspension mechanism, we are reducing the impact felt by the back area, reducing the load that the kid feels. Just like in a trophy truck, there's a vertical movement of the wheel, whilst the truck body remains unfazed by its motion. The bag entails a similar mechanism, wherein, the frame is attached to the body and the suspension links are attached to the backpack. The spring compensate for

IOP Conf. Series: Materials Science and Engineering 1194 (2021) 012006 doi:10.1088/1757-899X/1194/1/012006

the vertical motion of the body. For adjusting the spring damper, there is a worm gear drive which changes the spring stiffness as the bag is being loaded for optimal spring damping of the load. The major application of this design is in the Armed Forces, where heavy loads have to be carried out in extreme terrains, and stability is a major issue.

2. Equation Modelling

2.1. Body Motion

While designing the suspended-load pack, the vertical direction of body motion is considered because the pack is meant to oscillate along this direction. The motion of the body centre is often approximated by a sinusoidal profile as per (1) given below [2].

$$x = X \sin(\omega t) \tag{1}$$

Where X is the amplitude, ω is the radial frequency, and t is the time. Using this approach, the vertical movement can be validated using an inverted pendulum model with the straight leg acting as the oscillating body. This relates to the concept of the pendulum, where the upper body will oscillate once about the support leg before the other leg contacts the ground. A variant of this popular algorithm, as given in (1), resembles each step where the person's body will move from the lower trough on both the legs, to its upper crest on a single leg returns back to its initial lower trough. The design of the current study can be used to determine position of the body with the straight leg model. The key parameters, including the amplitude of oscillation may be summarised as follows

$$X = \frac{l_y}{2} \times 1 - \sqrt{1 - \frac{0.963v}{l_y \times 2 \times 1 \times 504 \times \frac{v}{l_y}^{0.57}}} \tag{2}$$

, wherein l_y and v denoting the length of the leg the trodding velocity. The leg length for an average Indian is approximated to be using the available anthropometric data [3].

2.2. Frame Motion

The main schematics of our backpack consist of three springs and two dampers. The frame associated with the body is connected to the backpack using spring and dampers [4]. Introduction of sinusoidal equation in the backpack trajectory undertakes a motion similar to that of the body. This was expressed as the following formula,

$$y = Y \sin(\omega t - \varphi) \tag{3}$$

Similar to (1), the Y represents the amplitude of the body and ϕ denotes the relative phase shift of the backpack with respect to the body motion [5]. The application of the spring stiffness k and damping coefficient c in the typical equation of motion for the backpack model is demonstrated below

$$k(x - y) + \frac{c}{2}(\dot{x} - \dot{y}) - m\ddot{y} = 0 \tag{4}$$

$$c = \frac{k_1 k_2}{k_1 + k_2} \tag{5}$$

Equation (4), when solved for a steady state response, provides the amplitude ratio in (6) and the phase shift in (7) as represented below,

$$\left|\frac{Y}{X}\right| = \left(\sqrt{\frac{k^2 + \omega^2 c^2}{(k - m\omega^2)^2 + \omega^2 c^2}}\right) \tag{6}$$

$$\varphi = a \tan\left(\frac{mc\omega^2}{\omega^2 c^2 + k^2 - m\omega^2 k}\right) \tag{7}$$

In addition, a static deflection in the spring will be observed due to the weight the empty backpack and load inside it, such that

$$\delta = \frac{mg}{k} \tag{8}$$

It should be noted that the he total elongation of the spring will comprise of the static deflection due to load and the relative phase shift in the positions of the body and the bag. The total force exerted by the backpack in the downward direction is equal to the sum of the weight acting on the frame and the force experienced by the backpack due to the motion. The following equation expresses the resulting frame force F_{frame} experienced by the body in the downward direction.

$$F_{frame} = mg - |F_{osc}| \sin(\omega t - \varphi) \quad (9)$$

In the equation derived above, the amount of the oscillating force $|F_{osc}|$ from the frame acting on the person can be calculated. Shown below is the expanded equation for oscillating force, which depends on the motion frequency, motion amplitude of the body, frame mass, stiffness, and damping.

$$|F_{osc}| = m\omega^2 X \sqrt{\frac{k^2+\omega^2 c^2}{(k-m\omega^2)^2+\omega^2 c^2}} \quad (10)$$

To obtain a minimal value for the oscillating force, the frame stiffness and damping coefficients have to be determined for a given frequency, frame mass, and the body's amplitude of oscillation [6]. When the damping of the system is removed, we obtain its resonance as shown.

$$k_{resonance} = m\omega^2 \quad (11)$$

It is important to define the stiffness value where the oscillating force will remain unaffected by the damping coefficient. This quantity is defined as the optimal stiffness k_{opt} , which is a function of the frame mass and motion frequency, as follows

$$k_{opt} = \frac{1}{2} m\omega^2 \quad (12)$$

This parameter, in accordance to the maximum phase shift is derived from ϕ, given in (8), with respect to the stiffness k and equating it to zero. It was observed that the maximum phase shift ϕ_{max} is obtained when reaching values of optimal stiffness. The maximum phase shift is expressed as follows,

$$\varphi_{max} = \text{atan}\left(\frac{mc\omega}{c^2-\frac{1}{2}m^2\omega^2}\right) \quad (13)$$

2.3. Stiffness adjustment

As the load on the backpack increases, there will be an increase in oscillation force for the same amplitude and frequency. To reduce the force, there needs to be a change in the model stiffness such that the oscillation force and phase shift remains constant. By adjusting the damper preload, the stiffness of the system can be changed. The addition of the damper preload will ensure that the values of the amplitude and the frequency have minimal deviation when there is a change in the backpack mass. Assuming a mass m_{opt}, which is the average mass for highest and the lowest loading capacity for the army backpack, the k_{opt} is decided. The following equation calculates the amount preload required

$$x_{pre} = \frac{m_1 g}{k_{opt}} - \frac{m_{opt} g}{k_{opt}} \quad (14)$$

Where m_l is the loaded weight and positive value specifies adding a preload and negative value specifies removing the preload. For applying preload to the spring, the shaft of the spring and its end plate has to be threaded. The relation between the minimum torque M_t required to add preload for the maximum load m_{max} is

$$M_t = \frac{m_{max} g d_m}{2} tan\,(\varphi - \alpha) \quad (15)$$

Where d_m is the mean diameter of the screw, ϕ is the friction angle and α is the lead angle [7]. A demarcation on the frame for the optimum load height will ensure the proper preload is set on the springs by the user. The preload should be adjusted every time there is a change in the loading.

2.4. Load distribution

For the sake of estimation of the load path, average values for the frame weight and backpack load are assumed. The weight of the frame and the backpack is assumed to be around 1.2 kg and 10 kg respectively. Hence for distribution of load, the areas of the body chosen are

- Sternum
- Shoulders
- Hips

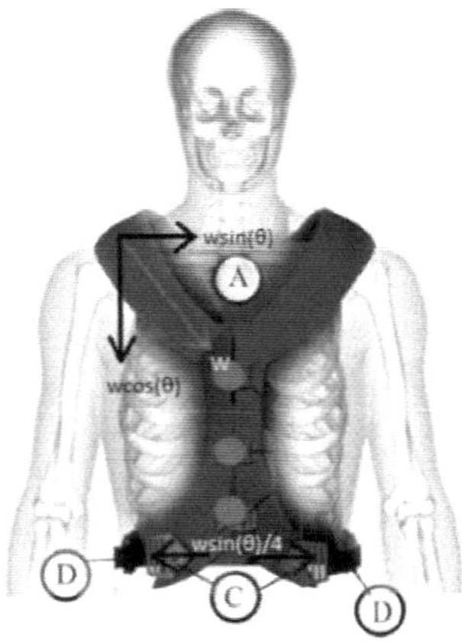

Figure 1. Force distribution on the joints of the pack

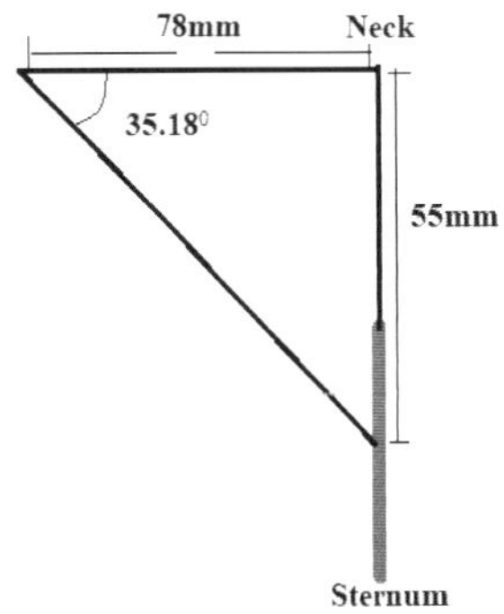

Figure 2. Calculation of strap angle θ

2.4.1. Sternum The sternum is the central bone of the ribs. The distribution of load on the sternum will greatly reduce the moment which is created by the bag since most of the load will be distributed on the sternum. In the figure 1, A is the frame which will house the spring and the bag. The joints B will transmit the vertical load on the shoulders horizontally on the sternum marked in red. The strap joint D will have the hip belt attached to it. The calculated weight distribution is estimated for 112.7 N. This load is now transmitted to the sternum as the tension at the shoulders is same as the tension at the joint. Hence the entire load of the pack is now transmitted to horizontally into the body, i.e. at the sternum.

2.4.2. Shoulders The weight of the pack is now distributed to the sternum; hence the weight on the shoulders will depend on the tension on the back-pack strap that the user has set. In the figure 2 given above, there are four plastic buckles which compensate the horizontal force at the shoulders. The weight W acts in the direction along the strap. The straps are then attached together at an angle θ , which is decided by the force rating of the buckles and the average midpoint of the shoulders. The weight is then distributed to w cos θ and w sin θ . The vertical component w cos θ is distributed to the shoulders and the horizontal component w sin θ is distributed to the three chest buckles and one joint which attaches the lower strap end to the bag. The buckles selected have a force rating of 20 kgs. Hence the straps will be able to take a force of 20/sin θ . A triangle, as shown in the figure below , gives the angle at which the straps have to be with respect to the body. The average distance between the centre of the neck and the midpoint of the shoulder is 78 mm. The average distance between the neck and the centre of the sternum is 55mm [8]. Hence the calculated angle is taken approximately to be 36°.

2.4.3. Hips Since the weight is distributed at the sternum using three buckles , the load on the hips is one fourth of the horizontal force on the strap . The hip strap are also used to balance the moment of the bag load. The ideal load on the sternum. The hips can carry maximum load hence it is used to carry the maximum load.

3. Design Analysis

To validate the working of the mechanism, component was designed in Solidworks. Figure 3 shows the basic design used for the testing. A W shaped frame was conceived for the sprung mass. This frame travels concentric to the main frame using linear motion bearings. The spring placed at the upper end and two dampers placed at the lower end were set for the suspension system. For testing the damping parameters of the frame, a test model was setup in MATLAB Simulink with the above calculated parameters. The setup is shown in the figure 4. After the above model was executed, the following graphs were obtained. A change in the damping amplitude is observed in figure 6 with reference to figure 5, and hence a decrease in the oscillation force is seen.

IOP Conf. Series: Materials Science and Engineering 1194 (2021) 012006 doi:10.1088/1757-899X/1194/1/012006

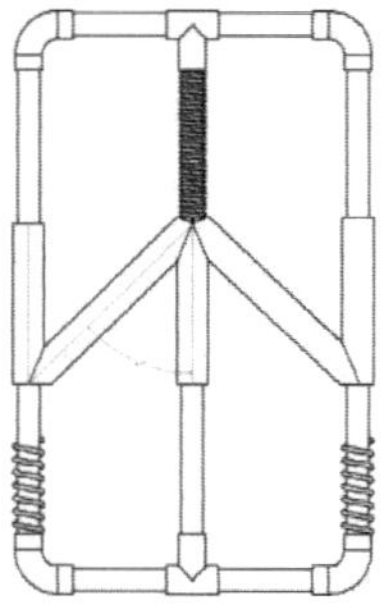

Figure 3. Proposed Frame design

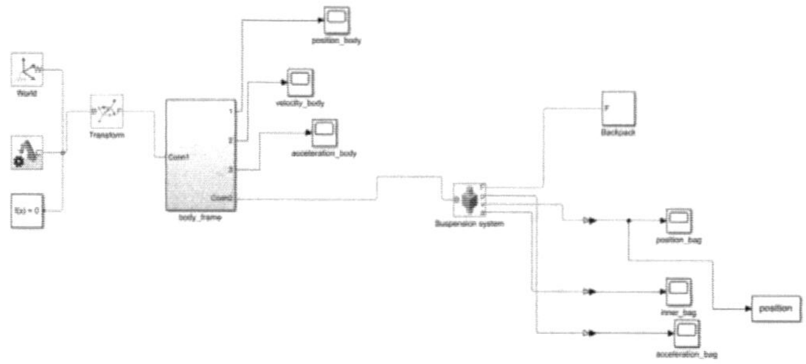

Figure 4. Suspended backpack Schematics in Simulink

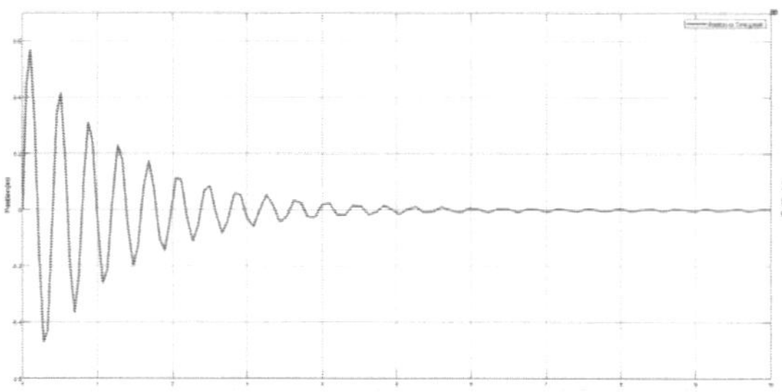

Figure 5. Acceleration vs time graph of the backpack without preload

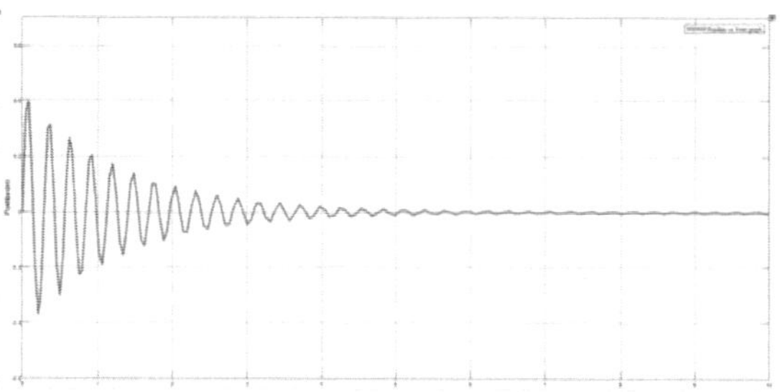

Figure 6. Acceleration vs time graph of the backpack with preload

4. Conclusion

The preload adjustment reduces the excess oscillation force experienced by various loads, thus preventing the sore shoulders and the misaligned spine. The redesign of the straps also helped correct bad posture and reduce the moment of the backpack about the body. The spring damper system greatly helped in reducing the shocks of the backpack when running or jumping, thus reducing the fatigue while long strenuous walk/runs.

5. References

[1] Publishing, Harvard. 2020. "Why Good Posture Matters - Harvard Health". *Harvard Health.*

[2] Saunders, J., Inman, V.T., Eberhart, H.D.: The major determinants in normal and pathological gait. *J. Bone Jt Surg* **35**, 543–558 (1953) .

[3] Roebuck, J.A., Kroemer, K.H.E., Thomson, W.G.: Engineering Anthropometry Methods. *Wiley-interscience*, New York (1975).

[4] Foissac, M., Millet, G., Geyssant, A., Freychat, P., Belli, A.: Characterization of the mechanical properties of backpacks and their influence on the energetic of walking. *J. Biomech.* **42**, 125–130 (2009).

[5] Rao, S.S.: Mechanical Vibrations. *Prentice Hall*, Singapore (2005)

[6] Rome, L., Flynn, L., Yoo, T.D.: Biomechanics: rubber bands reduce the cost of carrying loads.Nature 444, 1023–1024 (2006).

[7] Bhandari,V., 2017, Design of machine elements (New Delhi: *McGraw-Hill Education* (India)).

[8] Kim, M., Yoo, W., 2013, Effect of the Spacing of Backpack Shoulder Straps on Cervical Muscle Activity, Acromion and Scapular Position, and Upper Trapezius Pain, *Journal of Physical Therapy Science,* 25 685-686

IOP Conf. Series: Materials Science and Engineering 1194 (2021) 012007 doi:10.1088/1757-899X/1194/1/012007

Research of Ant Colony Algorithm with Elite Strategy in Process Route

FengYun Huang, ShiQiu Jiang

School of Mechanical and Electrical Engineering, Wuhan University of Technology, Wuhan, China

Email: 2747222643@qq.com

Abstract. The basic ant colony algorithm is slow in convergence and easy to fall into the local optimal solution. In this paper, it is proposed to apply the ant colony algorithm with elite strategy to the sequencing of the processing technology routes. With the least number of process equipment transformations and the shortest path of the tool, automatic ordering of the parts processing is realized. Simulation experiments show that the ant colony algorithm with elite strategy converges to the optimal solution after 39 iterations, while the basic ant colony algorithm converges after 71 iterations. Comparing the time-consuming of two algorithms, the ant colony algorithm with elite strategy improved by 42.7%. It illustrates that the ant colony algorithm with elite strategy has better application in sequencing process routes, which is beneficial for shortening the process planning time and increasing the process design efficiency.

1 Introduction

For many manufacturing companies, high-efficiency, low-cost, and high-quality production methods will have a significant impact on the survival, competition and development of enterprises. Therefore, efficiently formulating and generating optimal, low-cost and short-time processing routes are the focal points of scholars' research. Many scholars have been studying the decision-making and optimization of parts process routes to obtain low-cost, high-efficiency parts process routes that meet the actual production requirements. Zheng Yongqian[1] and others used genetic algorithms to select and sort the processing schemes, and verified that this method can improve the optimization ability of the processing technology; Pan Yuling[2] and others used the improved genetic algorithm to avoid the "premature" disadvantage of a single genetic algorithm; LI Xinyu[3] provided a blueprint for efficient manufacturing system. This paper proposes a novel algorithm hybridizing the genetic algorithm with strong global searching ability and variable neighborhood search with strong local searching ability for the IPPS problem; XY Li[4] proposed a flexible process optimization method based on genetic programming, and compared it to genetic algorithms which proved that the algorithm is adaptable and superior; Lan Xuan PHUNG[5] proposed a clustering algorithm applied to the CAPP process planning to generate the optimal operation sequence, which can effectively shorten the operation time; G. Nallakumarasamy[6] used simulated annealing technology to generate a feasible sequence of operations on the basis of the cost priority matrix and the reward and penalty matrix, and proved the feasibility and robustness of the algorithm; F. Zhang[7] proposed a new computer-aided process planning model. This method considers multiple decision-making activities and illustrates its performance through examples. Milica Petrovića[8] provided a new algorithm for optimization of flexible process plans based on utilization of particle swarm optimization algorithm and chaos theory, experimental results show that the developed method outperforms genetic algorithm, simulated annealing, hybrid GA-SA and generic PSO based approach; Su Yuliang[9] proposed an edge selection strategy based GA to solve the premature convergence problem may occur facing some complicated PCOSPs; Zhaohui Deng[10] proposed a multi-objective machining process route optimization model based on the genetic algorithms (GA), and the minimum processing time and the optimal carbon efficiency were set as the optimization objectives. Yujia Wu[11] formulated the priority matrix according to the sorting rules of the working steps, which makes the processing process more standardized and convenient. However, from the existing literature, many researches mainly optimizing the cost of machine conversion, tool conversion and fixture conversion when optimizing the process route, and consider the tool path less.

2 Introduction to related algorithms

2.1 Basic ant colony algorithm

Ant colony algorithm is one of many bionic algorithms. It simulates the behavior of ants foraging. In the process of foraging, ants will release pheromones on their paths. Ants have high probability to choose the path with high pheromone concentration and release a certain amount of pheromones to form a positive feedback mechanism, which makes the search results converge to approximate the optimal solution[12].

The most important thing in the ant colony algorithm is to determine the transfer rule of ants transferring from one point to another point, and the update rule of pheromone on the path. The state transition rule of ant k at time t in the basic ant colony algorithm is:

$$P_{ij}^{k}(t)=\frac{\tau_{ij}^{\alpha}(t)\eta_{ij}^{\beta}(t)}{\sum_{u\in N_k^i}\tau_{iu}^{k}(t)\eta_{iu}^{k}(t)} \qquad j\in N_k^i \tag{1}$$

In the formula, P_{ij}^{k}: the probability of ant k transferring from point *i* to point *j*; τ_{ij}: pheromone intensity on path l_{ij}; η_{ij}: heuristic information on path l_{ij}; α: heuristic factor; β: expected heuristic factor; N_k^i: The point that Ant *k* is allowed to visit next, $N_k^i = \{0,1,\dots u,\dots n-1\}$, n is the number of points. The heuristic information η_{ij} generally takes the reciprocal of the distance between points, $\eta_{ij} = 1/d_{ij}$, the greater the distance, the smaller the probability of being selected.

After completing a cycle, the ants need to adjust the pheromone on the path. The pheromone update principle of the general ant colony algorithm is:

$$\tau_{ij}(t+1)=(1-\rho)g\tau_{ij}(t)+\Delta\tau_{ij}(t,t+1) \tag{2}$$

$$\Delta\tau_{ij}(t,t+1)=\sum_{k=1}^{m}\Delta\tau_{ij}^{k}(t,t+1) \tag{3}$$

In the formula, $\Delta\tau_{ij}^{k}$: the amount of pheromone released by ant *k* on path l_{ij}; ρ: pheromone volatilization factor; $\Delta\tau_{ij}^{k}(t,t+1)$ represents the amount of pheromone left on the path l_{ij} of ant k at time $(t,t+1)$; $\Delta\tau_{ij}(t,t+1)$ is the pheromone increment on the path l_{ij} in this cycle.

2.2 Ant colony algorithm with elite strategy

Although the ant colony algorithm has a strong capability of global searching, it has a long search time and is prone to stagnation. Therefore, many improved algorithms have been proposed, and the ant colony algorithm with elite strategy is one of them. The core of the ant colony algorithm with elite strategy is to give the optimal solution obtained so far after each cycle with an extra amount of pheromone, so that in the next cycle the ants will be more concentrated near the optimal solution. In the ant colony algorithm with elite strategy, the updating principle of pheromone is different from the basic ant colony algorithm, as follows:

$$\tau_{ij}(t+1)=\rho g\tau_{ij}(t)+\Delta\tau_{ij}+\Delta\tau_{ij}^{*} \tag{4}$$

$$\Delta\tau_{ij}=\sum_{k=1}^{m}\Delta\tau_{ij}^{k} \tag{5}$$

$$\Delta\tau_{ij}^{k}=\begin{cases}Q/L_k & \text{The kth ant passes the path J again in this cycle}\\ 0 & \text{others}\end{cases} \tag{6}$$

$$\Delta\tau_{ij}^{*}=\begin{cases}eQ/L^* & \text{Path } l_{ij} \text{ is part of the optimal solution found}\\ 0 & \text{others}\end{cases} \tag{7}$$

In the formula, $\Delta\tau_{ij}^{*}$ represents the pheromone increment caused by elite ants on the path after walking through the path l_{ij}; e represents the number of elite ants; L^* represents the path length of the optimal solution found. The ant colony algorithm with elite strategy enables ants to find a better solution earlier in the algorithm process, and increase the efficiency of the algorithm.

3 Implementation of ant colony algorithm with elite strategy in process routing

Step1: Initialize the ant colony algorithm: initialize the number of iterations, and set the iteration counter NC = 0, the maximum number of iterations NC_max; initialize the pheromone on each path, set $\tau_{ij}(0) = C$, (*C* is a small constant), $\Delta\tau_{ij} = 0$; set the taboo table of each ant to $Tabu_k(k = 1, 2, \dots, m)$; initialize the priority matrix of the processing primitives;

Step2: Detect the priority matrix, determine the initial processing primitives according to the constraints of the priority matrix, place all ants randomly on these processing primitives, and update the ant's taboo table;

Step3: Ant *k* starts to find the optimal path;

Step4: According to the taboo table and the priority processing matrix, the ants determine the processing primitives that can be transferred. Then calculate the probability of the ants transferring to these processing primitives according to the state transition rules, and select the processing primitive u_j with the highest probability to transfer. Add u_j to the taboo table of ants $Tabu_k$; set all the elements of column *j* corresponding to u_j in the priority matrix to 0, and update the priority matrix;

Step5: $j = j + 1$; if $j < N$, go to Step4;

Step6: $k = k + 1$; if $k < m$, go to Step3;

Step7: Calculate the path length after each ant traverses the processing primitives, and compare with the current optimal path, if it is less than the optimal path, update the current optimal path;

Step8: According to the pheromone update rule with elite strategy ant colony algorithm, update the pheromone on the path;

Step9: NC = NC + 1; if NC <NC_max, clear the taboo table $Tabu_k$ of all ants, initialize the priority matrix of the processing primitives, go to Step2 otherwise, output the optimal path.

4 Results and analysis of two algorithms

In order to compare the results of the two algorithms, we now take the driving gear part of the automobile final reducer as an example. The two-dimensional diagram of this part is as shows in Figure 1:

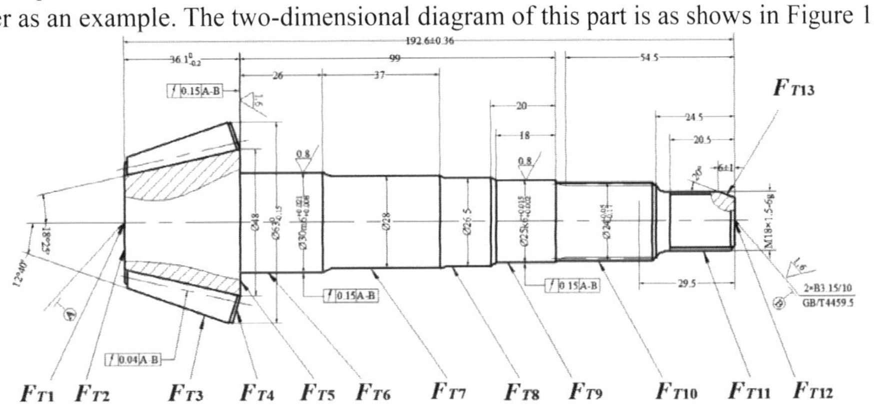

Figure 1. Two-dimensional drawing of driving gear of main reducer

According to the characteristic surface information, based on the constraints of the processing primitives, the information table of the processing primitives is obtained as shown in Table 1:

Table 1. Processing primitive information table.

u_i	Number	$u_i\_S_i$	$u_i\_M_i$	$u_i\_T_i$	$u_i\_C_i$	$u_i\_f_i$	$u_i\_r_i$
u_1	F_{T3}	1	1	1	1	F_{T1}, F_{T2}, F_{T13}	(16.5,0,0)
u_2	F_{T3}	2	2	2	2	F_{T5}, F_{T6}, F_{T9}	(16.5,0,0)
u_3	F_{T3}	3	2	3	2	F_{T5}, F_{T6}, F_{T9}	(16.5,0,0)
u_4	F_{T3}	4	3	4	3	F_{T6}, F_{T9}	(16.5,0,0)
u_5	F_{T4}	1	1	1	1	F_{T1}, F_{T2}, F_{T13}	(34.5,0,0)
u_6	F_{T5}	5	1	1	1	F_{T1}, F_{T2}, F_{T13}	(36,0,0)
u_7	F_{T5}	6	1	5	1	F_{T1}, F_{T2}, F_{T13}	(49,0,0)
u_8	F_{T6}	7	1	1	1	F_{T1}, F_{T3}, F_{T13}	(49,0,0)
u_9	F_{T7}	6	1	1	1	F_{T1}, F_{T2}, F_{T13}	(67.5,0,0)
u_{10}	F_{T8}	6	1	1	1	F_{T1}, F_{T2}, F_{T13}	(75.5,0,0)
u_{11}	F_{T9}	6	1	5	1	F_{T1}, F_{T2}, F_{T13}	(85.5,0,0)
u_{12}	F_{T9}	7	1	1	1	F_{T1}, F_{T3}, F_{T13}	(85.5,0,0)
u_{13}	F_{T10}	6	1	1	1	F_{T1}, F_{T2}, F_{T13}	(105,0,0)
u_{14}	F_{T10}	8	4	6	1	F_{T1}, F_{T13}	(105,0,0)
u_{15}	F_{T11}	6	1	1	1	F_{T1}, F_{T2}, F_{T13}	(180,0,0)
u_{16}	F_{T11}	9	4	7	1	F_{T1}, F_{T13}	(180,0,0)
u_{17}	F_{T13}	10	5	8	4	F_{T2}, F_{T6}, F_{T9}	(180,0,8)

In the above table, $u_i\_s_i$: the machining method used for machining primitives, 1 means fine turning cone, 2 means rough milling gear, 3 means fine milling gear, 4 means grinding gear, 5 means fine turning plane, 6 means fine turning the outer cylindrical surface, 7 means fine turning after heating, 8 means splining, 9 means rolling thread, 10 means milling bevel; $u_i\_M_i$: the machine tool used for machining primitives, 1 means external cylindrical lathe, 2 means Gleason gear milling machine, 3 means special grinding machine, 4 means horizontal gear rubbing machine, 5 means horizontal milling machine; $u_i\_T_i$: the tool used for machining primitives, 1 means fine turning outer cylindrical turning tool, 2 means rough milling gear tool, 3 means fine milling gear tool, 4 means grinding sand, 5 means fine cylindrical turning tool for bear, 6 means tool for rolling threads, 7 means tool for rolling splines, 8 means disc milling tool; $u_i\_C_i$: the fixture used for machining primitives, 1 means the special center chuck, 2 means the fixture for special milling gear, 3 means the fixture for special grinding gear, 4 means the fixture for special vertical milling; $u_i\_f_i$: the clamping surface in machining primitive machining; $u_i\_r_i$: the center coordinates of the feature surface to which the processing primitive belongs.

In the processing, if the machine tools, tools, fixtures, etc. of the two processing elements matches, they need to be arranged processing together to reduce the change of process equipment. Therefore, clustering constraint relationships between the processing elements need to be taken into sufficient consideration when defining "distance". At the same time, it is also necessary to consider the path of the tool. When there are multiple processing primitives with the same process equipment, the spatial distance between the processing primitives needs to be considered. The processing primitive with a shorter spatial distance is preferentially processed first to reduce the moving distance of the tool, instead of randomly selecting these machining primitives. For example, when processing shaft parts, the tool will process the surface of the part in the same direction in one process instead of reciprocating. The distance between processing primitives can be defined as:

$$\begin{aligned} D(u_i,u_j) = {} & \omega_s g sim(u_i\_s_i,u_j\_s_j)+\omega_M g sim(u_i\_M_i,u_j\_M_j)+ \\ & \omega_T g sim(u_i\_T_i,u_j\_T_j)+0.001 lg d(u_i\_r_i,u_j\_r_j)+ \\ & (\omega_C+\omega_f) g \max[sim(u_i\_C_i,u_j\_C_j), sim(u_i\_f_i,u_j\_f_j)] \end{aligned} \quad (8)$$

The values of the weighting coefficient ω in the formula are: ω_S: 0.3, machine tool ω_M: 0.3, tool ω_T: 0.15, fixture ω_C: 0.15, clamping surface ω_f: 0.1, $sim(a,b)$ is a similarity judgment function, $d(u_i\_r_i, u_j\_r_j)$ is the distance function between the processing primitives, $max[sim(u_i\_C_i, u_j\_C_j), sim(u_i\_f_i, u_j\_f_j)]$ means to consider the transformation of both the fixture or the clamping surface of the processing primitive , that is, take the maximum value.

$$sim(a,b)=\begin{cases}1 & a\neq b\\ 0 & a=b\end{cases} \quad (9)$$

$$d\left(u_i\_r_i,u_j\_r_j\right)=0.001\lg\sqrt{\left(x_i-x_j\right)^2+\left(y_i-y_j\right)^2+\left(z_i-z_j\right)^2} \quad (10)$$

In this paper, the goal is to minimize the number of process equipment changes and the shortest tool path, and realize the automatic sequencing of parts processing procedures. According to the flow of ant colony algorithm and algorithm implementation, based on Matlab programming, the basic ant colony algorithm and the ant colony algorithm with elite strategy are used to realize the process sequencing of processing primitives. According to the research results of Xu Hongmei[13], the value of the number of ants m should be [0.6n, 0.9n], n is the number of cities in the TSP, so this paper takes m = 15 and the number of elite ants e = 4. When the heuristic factor $\alpha \in [1.0,\ 2.0]$ and the expected heuristic factor $\beta \in [3.0,\ 6.0]$, the performance of the ant colony algorithm is better. In this paper, α = 2 and β = 3. The larger the value of the pheromone constant Q, the worse the global search ability of the algorithm, and the computational power is very unstable. In this paper, Q = 1. The pheromone volatilization factor ρ is too small, which is not conducive to the accumulation of pheromone. It is value is too large, and it is easy to fall into the local optimal solution. Generally, ρ is [0.3, 0.9], and ρ = 0.4 in this paper. NC_max =100.

The processing primitive results obtained by using two algorithms are shown in figure 2 and figure 3. From the figure, it can be seen that the results obtained by the two algorithms are consistent, and the results meet the actual processing results. The results are as follows:

u_1-u_5-u_6-u_9-u_{10}-u_{13}-u_{15}-u_7-u_{11}-u_{14}-u_{16}-u_{17}-u_2-u_3-u_8-u_{12}-u_4。

```
ACO_Shortest_Route =

   1   5   6   9   10   13   15   7   11   14   16   17   2   3   8   12   4
```

Figure 2. Sorting results of basic ant colony algorithm.

```
ASelite_Shortest_Route =

   1   5   6   9   10   13   15   7   11   14   16   17   2   3   8   12   4
```

Figure 3. Sorting results of ant colony algorithm with elite strategy

The iteration curves of the two algorithms are shown in figure 4 and figure 5:

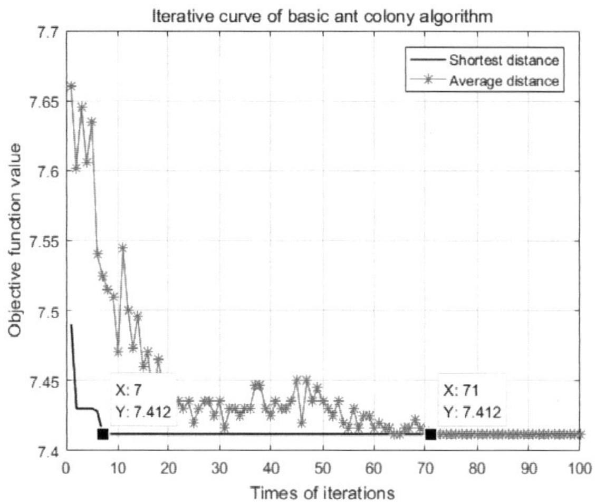

Figure 4. Iterative curve of basic ant colony algorithm

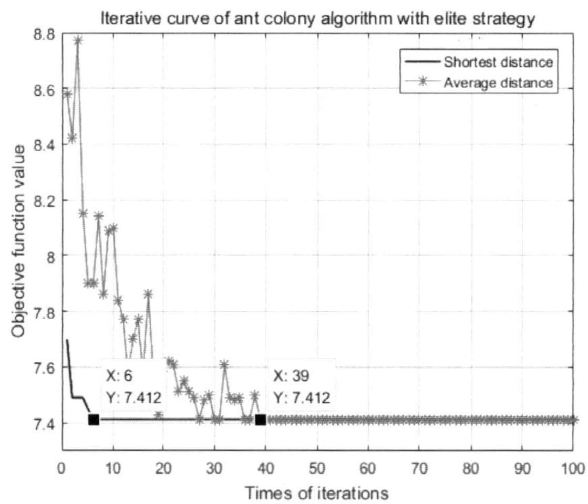

Figure 5. Iteration curve of ant colony algorithm with elite strategy

It can be seen from the figure 4 and figure 5, during the seventh iteration of the basic ant colony algorithm, some ants found the optimal solution with the shortest distance of 7.412; after the 71st iteration, the ants concentrated on the optimal solution and the algorithm converged. In the ant colony algorithm with elite strategy, during the sixth iteration, some ants found the optimal solution with the shortest distance of 7.412; after the 39th iteration, the ants concentrated on the optimal solution and the algorithm converged. Comparing the time consumption of the two algorithms, the basic ant colony algorithm takes 25.825s, and the elite ant colony algorithm takes 14.793s.

5 Conclusion

Using basic ant colony algorithm, the convergence speed is slow, and the ability to find the optimal solution is weak. Compared with the basic ant colony algorithm, the ant colony algorithm with elite strategy algorithm performs better in convergence speed and the ability to find the optimal solution. In addition, the ant colony algorithm with elite strategy takes less time to make processing element sequencing decisions, and is more suitable for solving processing element sequencing problems, which is beneficial for shortening process planning time and increasing process design efficiency.

6 References

[1] Yongqian Z. Optimization of Process Selection and Sequencing Based on Genetic Algorithm[J]. *China Mechanical Engineering*, **23**: 59-65, 2012.

[2] Yu-ling P. Optimization of Process Route for Complicated Box Parts Based on Hybrid Genetic Algorithm[J]. *Modular Machine Tool & Automatic Manufacturing Technique*, **2**(2): 103-107, 2019.

[3] Li X., Gao L., Pan Q., et al. An Effective Hybrid Genetic Algorithm and Variable Neighborhood Search for Integrated Process Planning and Scheduling in a Packaging Machine Workshop[J]. *IEEE Transactions on Systems, Man, and Cybernetics: Systems*, **49**(10): 1933-1945, 2019.

[4] Li X.Y., Shao X.Y., Gao L. Optimization of flexible process planning by genetic programming[J]. *The International Journal of Advanced Manufacturing Technology*, **38**(1): 143-153, 2008.

[5] Phung L.X., Tran D.V., Hoang S.V., et al. Effective method of operation sequence optimization in CAPP based on modified clustering algorithm[J]. *Journal of Advanced Mechanical Design, Systems, and Manufacturing*, **11**(1), 2017

[6] Nallakumarasamy G., Srinivasan P.S.S., Venkatesh Raja K., et al. Optimization of operation sequencing in CAPP using simulated annealing technique (SAT)[J]. *The International Journal of Advanced Manufacturing Technology*, **54**(5): 721-728, 2011.

[7] Zhang F., Zhang Y.F., Nee A.Y.C. Using genetic algorithms in process planning for job shop machining[J]. *IEEE Transactions on Evolutionary Computation*, **1**(4): 278-289, 1997.

[8] Petrović M., Mitić M., Vuković N., et al. Chaotic particle swarm optimization algorithm for flexible process planning[J]. *The International Journal of Advanced Manufacturing Technology*, **85**(9): 2535-2555, 2016.

[9] Su Y., Chu X., Chen D., et al. A genetic algorithm for operation sequencing in CAPP using edge selection based encoding strategy[J]. *Journal of Intelligent Manufacturing*, **29**(2): 313-332, 2018.

[10] Deng Z., Lv L., Huang W., et al. A High Efficiency and Low Carbon Oriented Machining Process Route Optimization Model and Its Application[J]. *International Journal of Precision Engineering and Manufacturing-Green Technology*, **6**(1): 23-41, 2019.

[11] Wu Y.-j., Cao Y., Wang Q.-f. Assembly sequence planning method based on particle swarm algorithm[J]. *Cluster Computing*, **22**(1): 835-846, 2019.

[12] Songtao W. Improved genetic algorithm based on optimized genetic algorithm ant colony algorithm AGV path planning[J]. *Automation Application*, **3**(3): 47-49, 2017.

[13] Hong-mei X. The research on the parameters of the ant colony algorithm[J]. *Journal of ShanDong University of Technology(Natural of Science Edition)*, **22**(1): 7-11, 2008.

The Effect of Bank Angle on the Pressure Distribution and Deflection of Non-pneumatic Tires

R Sriwijaya, K Tjandra, and R Hamzah

Department of Mechanical and Industrial Engineering, Faculty of Engineering, Universitas Gadjah Mada, Grafika Street, Yogyakarta 55281, Indonesia

E-mail: sriwijaya@gadjahmada.edu

Abstract. A certain amount of road angle is necessary to reduce the risk of run-off-road and head-on crashes. Following the rapid development of non-pneumatic tires (NPT) in the past years, pressure distribution and deflection analysis of NPT while operating on flat and inclined road angles are analyzed using the Finite Element Method. The results showed that a higher stress concentration of spokes is evident at an elevated bank angle. Simultaneously, a more significant load combined with bank angle increment resulted in a higher displacement magnitude. This study aims to enable the designer to create an NPT design that could adapt to different bank angles.

1. Introduction

Pneumatic tires are currently the most popular type of tires available in the market. The fact that they can withstand contact pressure and provide buffering function due to internal inflation pressure supplied by air serves as the primary reason for its popularity. However, the dependence on air means that this type of tire requires air pressure maintenance and is susceptible to puncture. The disadvantages lead to the development of non-pneumatic tires (NPT), a single unit comprised of a rigid hub, deformable polyurethane spokes, and a tread band, by several tire manufacturers worldwide.

Figure1. Vehicle operation under certain road bank angle.

Several works have been carried out to explore different aspects of NPT. Sassi et al. [1] examined the design aspects of NPT and imply that good radial stiffness can be achieved by introducing cuts in tire design. Another research conducted by Aboul-Yazid et al. and Mohan et al. [2][3] analyzed rolling resistance, contact pressure, von Mises stresses, and mass value aspects to determine the characteristics performance of 4 different spokes structures. Zmuda et al. [4] developed a numerical model of NPT and was further validated by experiments. More specific research on NPT's spokes was conducted by Ju et al. [5] and Mathew et al. [6] by varying honeycomb spokes design and introducing diamond structure.

However, since tire cells dynamically engage with the road surface, it is crucial to consider the stress distribution and deflection of NPT. Two previous research in this area has been published by Sriwijaya and Hamzah. The former [7] suggests that tension deviation occurs in NPT, especially when applied in high velocity, while the latter [8] [9] concludes that road inclination angle increment will be followed by increases in pressure distribution, deflection, and the tendency of spoke failure.

Road bank angle, a phenomenon where the outer edges of curved are raised above the inner edge to provide sufficient centripetal force as shown in figure 1, is commonly applied in highway construction so vehicles can make a safe turn. An extreme case of road bank angle utilization can be seen on velodrome construction. By considering road topography, this research aimed to better understand NPT's pressure distribution and deflection under three different bank angles. Finite Element Analysis (FEA) using Abaqus® software is chosen to evaluate the environment due to its effective yet inexpensive method.

2. Modeling

2.1. Non-pneumatic tires

A 3D model of NPT made by Sriwijaya and Hamzah (figure 2) is referred for this research. The design utilizes honeycomb structures and consists of 3 layers. The inner and outer ring are made of polyurethane (PU) and connected by honeycomb spokes, while high-strength steel (HSS) is selected for ring hub material. Treads are attached to the outer rubber layer. Relevant properties of each material are provided in table 1.

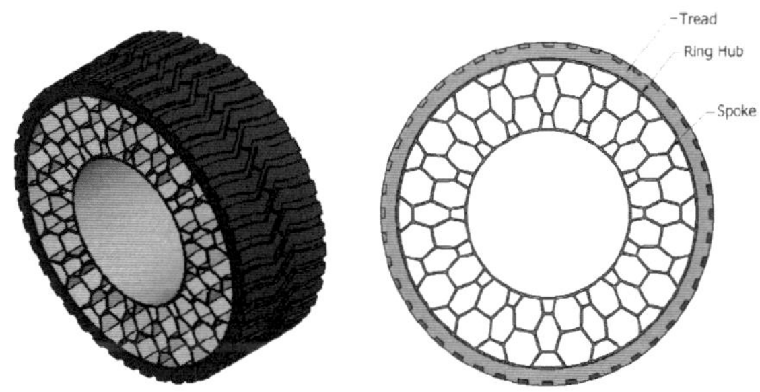

Figure 2. Airless tire geometry and components [8][9].

2.2. Road topography

This paper analyzes three different bank angles (0°, 1.8°, and 2.7°) in this paper to simulate practical applications. Illustrations of the tire under each bank angle are provided in figure 3.

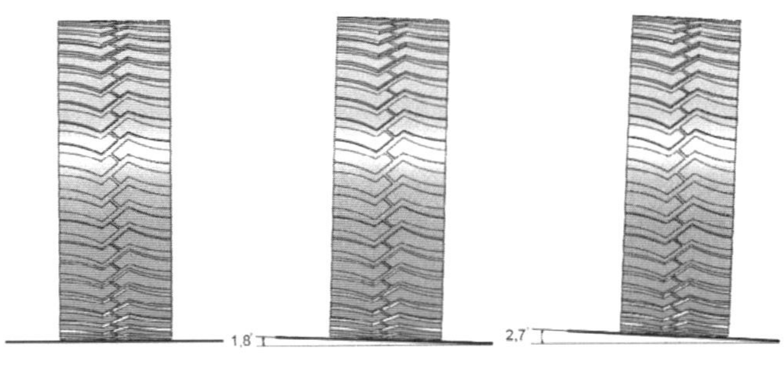

Figure 3. Tire models under different bank angle.

3. Finite element analysis

The NPT model is imported to FEA software. Abaqus®/ CAE 6.11 software is chosen to carry out the structural analysis. The numerical research was conducted for five load conditions: 1, 2, 3, 4, and 5 kN. Bank angles were also varied at 0°, 1.8°, and 2.7° during the simulation.

Table 1. Material properties of HSS, PU, and rubber.

Properties	AISI 4340 Steel	PU	Rubber
Density (kg/m^3)	7,800.00	1,200.00	1,150.00
Young's Modulus (GPa)	210.00	32.00	11.90
Yield Stress (Mpa)	1,240.00	145.00	16.00
Poisson's Ratio	0.29	0.49	0.49

4. Findings and discussion

Figures 3 to 6 show the results of static analysis of von Mises stress distribution on three different bank angles under the highest load (5 kN). The analysis was performed on the overall NPT system, shown by figure 4(a), figure 5(a), and figure 6(a). A separate analysis of von Mises stress was also carried out for both NPT's tread shown by figure 4(b), figure 5(b), figure 6(b), and NPT's spokes demonstrated by figure 4(c), figure 5(c) and figure 6(c).

Table 2. Yield stress of NPT's assembly, tread, and spoke.

	Assembly	Tread	Spoke
Yield stress (MPa)	710	145	16

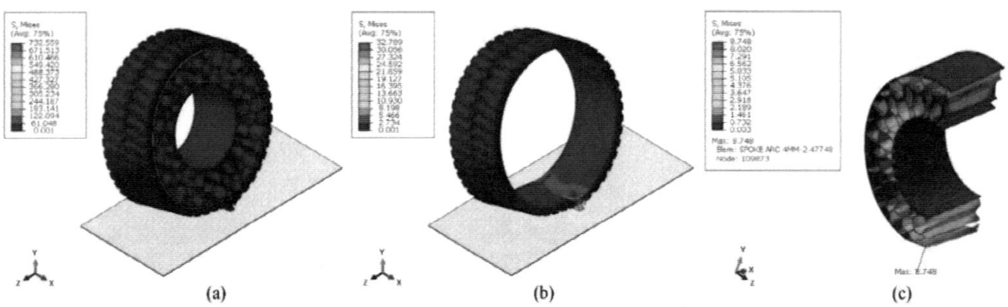

Figure 4. Von Mises stress distribution of (a) NPT, (b) tread, and (c) spokes under 0° bank angle.

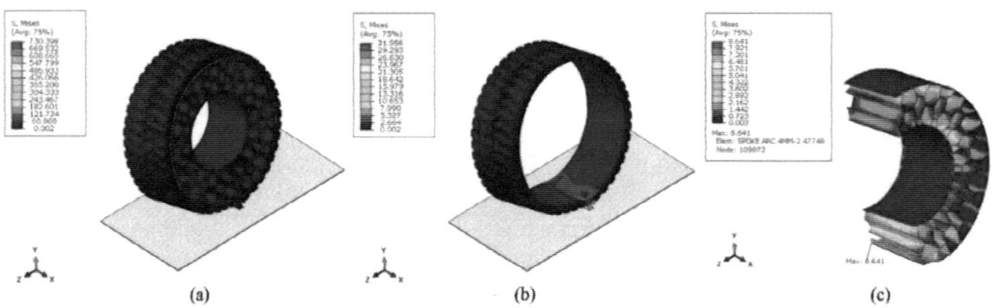

Figure 5. Von Mises stress distribution of (a) NPT, (b) tread, and (c) spokes under 1,8° bank angle.

IOP Conf. Series: Materials Science and Engineering 1194 (2021) 012008 doi:10.1088/1757-899X/1194/1/012008

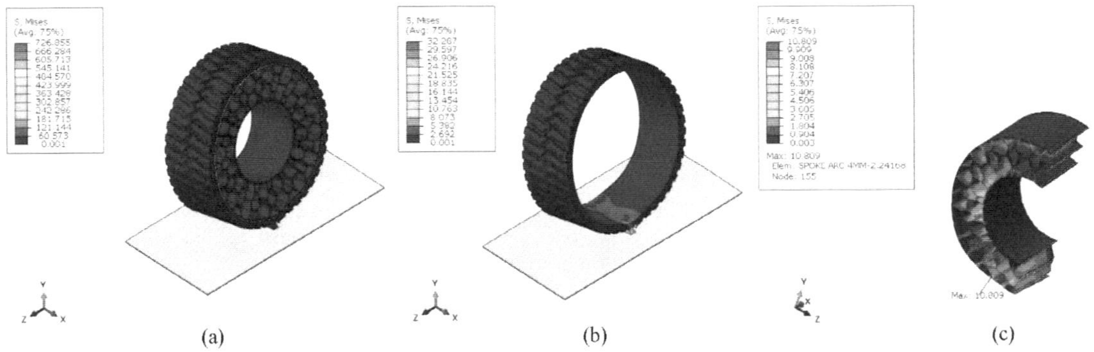

Figure 6. Von Mises stress distribution of (a) NPT, (b) tread, and (c) spokes under 2,7° bank angle.

Based on the results obtained, it can be inferred that stresses were well distributed around the overall NPT system and NPT's tread, although slightly higher tension was observed at the bottom part of the structures. However, NPT's spokes analysis results showed that significantly higher von Mises pressure distribution occurred at the spokes' joint area, which indicated the area with high-pressure concentration. Maximum von Mises stress of spokes are 8.748 MPa for 0° angle, 8.641 MPa for 1,8° angle, and 10.809 MPa for 2,7° angle. Von Mises stress of different load and bank angle induced in system assembly, tread, and spokes are shown in figure 6, while yield stress of each component is presented in table 2. Based on the visual observation of the simulation, it is also suggested that the deformation of spokes is higher at the lower part of the NPT.

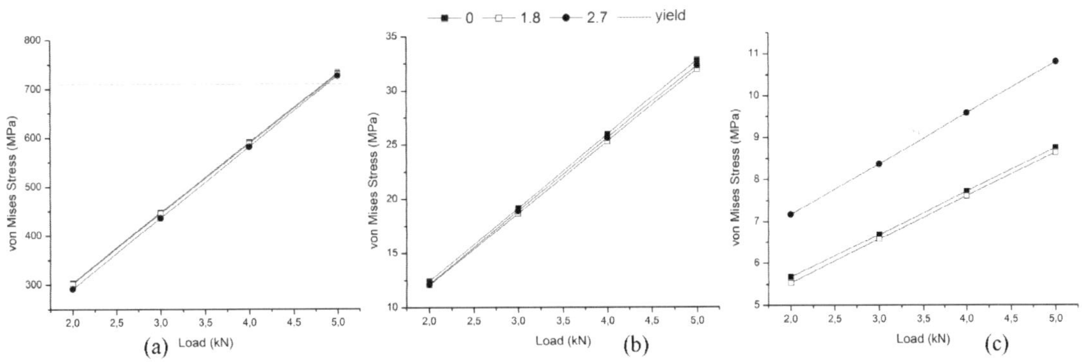

Figure 7. Von Mises maximum stress of (a) NPT (b) tread and (c) spokes on different bank angles.

Maximum stress obtained at different load and bank angles is recorded and plotted into line graphs, as shown in figure 7. It is clear that both tread and spokes can withstand maximum stress with quite a large margin within the experimented load. However, the pressure absorbed reached its yield strength when 5 kN of load was applied to the system, with the highest stress concentration shown at the ring hub. Materials with the highest yield stress determine the yield stress property for each analysis. Therefore, yield stress for the entire NPT system is obtained from AISI 4340 steel with a yield stress of 710 MPa.

Each graph indicated that the increase of load is always followed by a linear increase in stress distribution, either for NPT's system, tread, or spokes. Consequently, the highest stress is always produced at 5 kN load.

The effect of bank angle on stress distribution can also be inferred from figure 7. An increase of bank angle complemented with higher load give the most significant increase in stress concentration at spokes. On the contrary, the amount of stress reduced at a higher bank angle while the whole system was investigated. Bank angle gave negligible influence towards the stress concentration of NPT's tread.

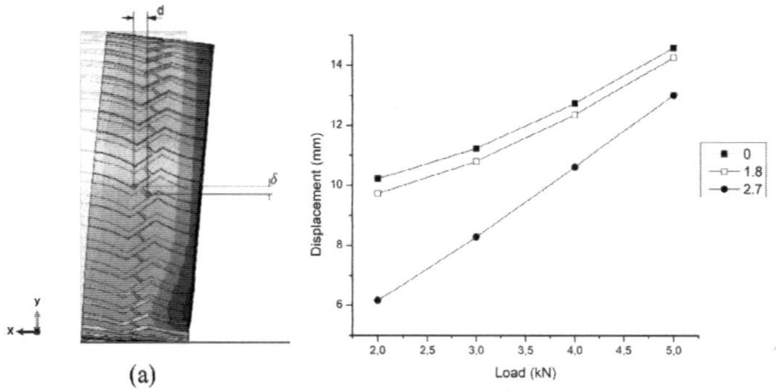

Figure 8. (a) Displacement measurement method; (b) NPT's displacement at different loads.

Displacement of NPT's data is obtained by measuring the shifted center of gravity at each axis (x, y, z) and illustrated by figure 8(a). Two red dots are utilized to represent each center of gravity.
The magnitude of displacement rises exponentially with the increasing load given to the NPT. It is evident that although the displacement of NPT at 0° bank angle is significantly higher than inclined angles, the gap quickly narrowed down as higher load applied to the system.

5. Conclusions

The increment of road bank angle supplemented with a high load applied to NPT resulted in increasing magnitude of both displacement and von Mises stress. In contrast to the NPT's thread, NPT spokes are determined to be the most vulnerable part of coping with inclined bank angle. It is also vital to notice that the bank angle's inclination gave more influence on NPT's spokes than other parts of NPT. Authors recognized that further research by taking more variables into account is required to simulate real-world terrain better.

6. References

[1] Sassi S, Ebrahemi M, Mozien M A and Hadary Y E 2016 *Int. J. Mech. Syst. Eng.* **2** 114
[2] Aboul-Yazid A M, Emam M A A, Shaaban S and El-Nashar M A 2015 *Int. J. of Automot. & Mech. Eng.* **11** 2212-2223
[3] Mohan A, Johny C A, Tamilarasu A, Bhasker J P and Ravi K 2017 *IOP Conf. Ser.: Mater. Sci. Eng.* **263** 062061
[4] Zmuda M, Jackowski J and Hryciow Z 2019, *AIP Conf. Proc.* **2078** 020027
[5] Ju J, Kim D and Kim K 2012 *Compos. Struct.* **94** 2285-2295
[6] Mathew N J, Sahoo D K and Chakravarthy E M 2017 *IOP Conf. Ser.: Mater. Sci. Eng.* **197** 012042
[7] Hamzah R and Sriwijaya R 2019 *5th Int. Conf. Sci. & Technol. (Yogyakarta)*
[8] Sriwijaya R and Hamzah R 2019 *AIP Conf. Proc.* **2187**, 050021
[9] Hamzah R 2019 *Master's Thesis Universitas Gadjah Mada*

Acknowledgments

The research was funded by Universitas Gadjah Mada through the RTA Grant No. 2129/UNI/DITLIT/DIT-LIT/LT/2019.

An Overview of NVH Problems Caused by Changes in the Excitation Source of HEV

Yi-fan CHEN, Hui-hua FENG*

School of Mechanical Engineering, Beijing Institute of Technology, Beijing100081, China

Email: fenghh@bit.edu.cn

Abstract. HEV (Hybrid Electric Vehicle) is the most promising vehicle among new energy vehicles. In view of the structuralchanges of hybrid electric vehicles compared with traditional fuel vehicles and the subsequent changes in NVH (Noise, Vibration, Harshness) performance, the main reasons for the vibration and noise of hybrid electric vehicles and the research results of NVH at home and abroad are detailed introduction. Through the classification and summary of the factors affecting the vibration and noise inside and outside the HEV, the advantages and disadvantages of the NVH control technology currently being developed and developed are summarized, and the future development of the NVH of the HEV is prospected.

1. Introduction

HEV (Hybird Electric Vehicle) is a new energy vehicle composed of two or more drive systems that can operate independently or at the same time. Because of its low fuel consumption and low emission characteristics, which can meet the current increasingly stringent vehicle emission regulations, HEV is a new energy vehicle type that has successfully achieved industrialization and has promotion significance 错误!未找到引用源。[2]. NVH (Noise, Vibration, Harshness) performance is one of the most important indicators to measure the quality of automobiles, and it has been paid more and more attention in design and manufacturing[3]. Compared with fuel vehicles, the structures of HEV have undergone several significant changes: the powertrain system integrates the engine, driving machine, engine start-stop motor, power coupling mechanism, final drive and other components, and can realize pure electric operation, engine drive, and hybrid electric drive. By increasing the energy-to-electric conversion system and related components, the energy utilization rate of the HEV is improved[4][5]. The NVH performance of HEV has changed greatly due to the electric motors and the power battery packs that are large in size and weight. Accompanied by the excitation of multiple complex power sources, the torsional vibration problem of the transmission system is more prominent, and the torque fluctuations during the switching of various driving modes can also bring sudden impacts, which have a great impact on the ride comfort of the entire vehicle[6]. Advanced electronic wire control technology is widely used in HEV, and the radiated noise generated by the booster circuit system becomes the noise source of high frequency noise inside and outside the HEV. In addition, due to the reduced shielding effect of engine noise, NVH problems that are not considered in traditional fuel vehicles, such as wind noise and road noise, have gradually become prominent.

2. Classification of HEV

Compared with traditional fuel vehicles, the structural changes of HEV are mainly manifested as follows[7]错误!未找到引用源。:1.the powertrain of HEV that integrates engine, drive motor, engine start-stop motor, gearbox, and final drive can realize pure electric driving, engine driving, and hybrid driving;

2.the driving mode of the auxiliary system is changed from engine-powered to electric-driven, which can ensure the orderly operation of air-conditioning compressor system, cooling system, vacuum assist system and other auxiliary systems under various working conditions. For this purpose, various electric pump structures must be added to HEV; 3.the adjustment of the power system and auxiliary system has caused changes in the quality, overall layout, load distribution and driving dynamics of the HEV. According to different power transmission methods, HEV can be divided into three types: series, parallel and parallel-series[9]. The composition differences of the three types of HEV are shown in the following Figure1-3[10]:

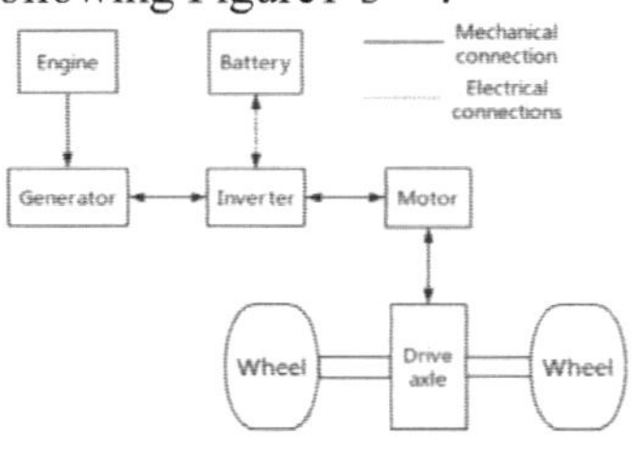

Figure 1. Series HEV

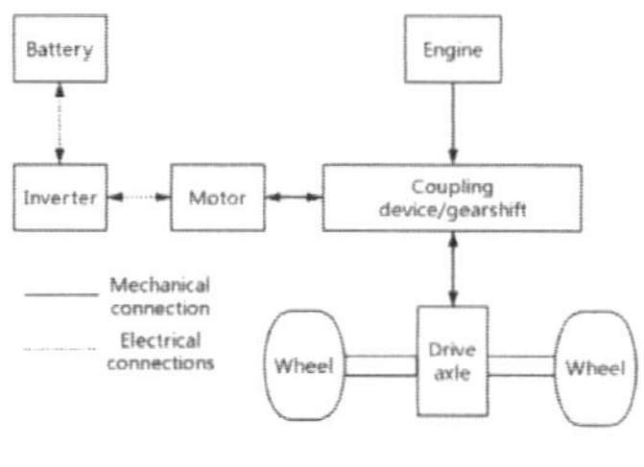

Figure 2. Parallel HEV

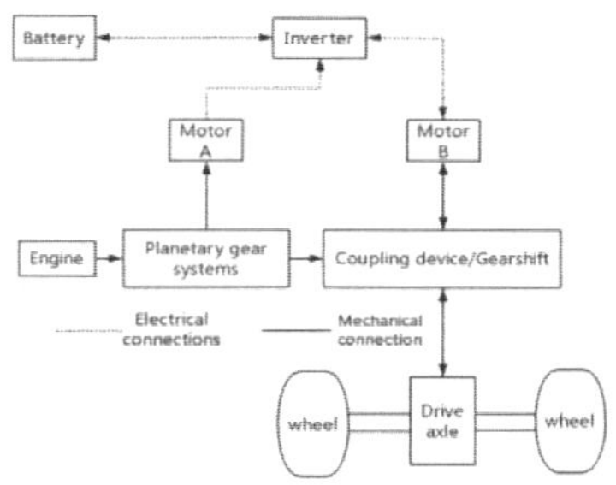

Figure 3. Parallel-series HEV

3. NVH Challenges Caused by Changes in Incentive Sources

3.1. Vibration and Noise of Motors

The noise frequency of the motor is often in the noise-sensitive frequency band of the human ears, so the NVH performance of the vehicle is largely affected by the motor noise. The NVH problems of motors can be divided into three types according to the different generation mechanisms[11][12]:electromagnetic vibration and noise, aerodynamic vibration and noise, and mechanical vibration and noise. Their causes and analysis are shown in Table 1.

Table 1. Classification of motor vibration and noise[13][14]

Types	Causes
Electromagnetic vibration&noise	Air gap magnetic waves can generate radial and tangential electromagnetic force that vibrate and deform the stator core and tooth roots. Electromagnetic alternating changes cause vibration and noise
Aerodynamic vibration&noise	The local air pressure in the ventilation system changes rapidly and pulsates sharply with time, and the friction between the gas and the air duct generates vibration
Mechanical vibration&noise	The vibration and noise caused by the interaction of moving parts of the motor during operation. The method of assembly and the processing materials of the components are all influencing factors

Since most modern motors use high-damping and low-noise sliding bearings or high-precision rolling bearings, and the stator and rotor have received strict dynamic balance correction, mechanical vibration and noise problems are not serious. Aerodynamic vibration and noise are not prominent when the motor speed is not high, and the aerodynamic noise of the water-cooled drive motor can basically be ignored. Therefore, electromagnetic noise is the main concern. The vibration and noise of the motor are mainly obtained by combining simulation analysis and experimental verification, and the simulation model is verified and optimized through experiments. The NVH optimization process of the motor includes three parts: structural modal calculation and experimental verification, forced vibration response calculation and experimental verification, and acoustic radiation calculation and verification, as shown in Figure 4. The optimization of the motor NVH performance mainly revolves around electromagnetic noise. The optimization scheme mainly includes electromagnetic excitation source optimization and motor structure optimization. By optimizing the modal of the motor housing, the optimization method to avoid the resonance of the motor structure caused by the excitation frequency of the electromagnetic force wave being the same as the modal frequency of the motor is relatively passive, poor in versatility, and does not have the significance of promotion.

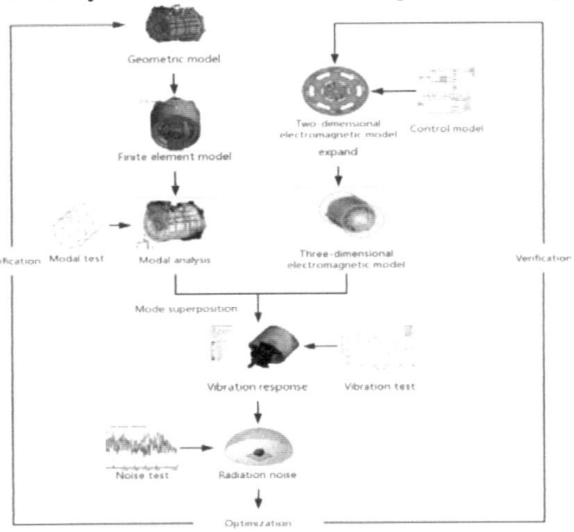

Figure 4. Simulation and prediction process of electromagnetic noise[15][16]

Therefore, the optimization of the electromagnetic noise of the motor is mainly from Start by optimizing the electromagnetic excitation source. The optimization of electromagnetic excitation sources can be divided into two directions:①Optimize the electromagnetic parameters: reduce the ellipticity of the stator inner circle and the rotor outer circle in the motor manufacturing process, while ensuring that the stator and rotor are concentric, so that the air gap can be uniform. In addition, noise can be reduced by reducing the air gap flux density and and increasing the air gap. but these may bring adverse effects on motor efficiency, power, and torque performance. Salon S[17], WANG[18], Hur J[19]and others adopt different optimization schemes for the rotor structure and winding arrangement, which effectively reduces the torque ripple, vibration and noise of the motor. ②Optimization of the control strategy: inject the compensation current to weaken the amplitude of the target electromagnetic force, appropriately increase the carrier frequency to improve the current waveform and reduce the relative noise. CHAI[20], LIU[21], Gwak[22], Taniguchi[23] and others improve the NVH characteristics of motors by designing harmonic synthesis, spectrum adjustment and voltage injection schemes.

Problems to be explored and studied: ①When the motor vibration and noise modeling is in the design stage, establish accurate stator structure models without the help of experiments; ②In addition to the model under steady-state operating conditions, establish a motor noise analysis model under non-steady-state operating conditions such as starting, acceleration and braking; ③Solve the problem that some solutions reduce motor noise while sacrificing motor performance.

3.2. Vibration and Noise of Motors

3.2.1. The NVH Challenge of HEV Transmission. The vibration of the transmission is a very complicated random vibration process. The rotation vibration of the shaft, the vibration of the gear transmission system and the high frequency vibration of the bearing are transmitted to the transmission case, causing the case to generate extremely complex random vibration. Various excitation forces are the vibration source of the complex vibration of the gearbox, and the noise of the gearbox mainly comes from the gear system[24]. The series HEV is not equipped with a gearbox mechanism, so the NVH performance of the transmission system can be improved. The noise of parallel and series-parallel HEV transmission mainly includes gear whistling and gear knocking[25]. The main cause of gear knocking noise is the obvious fluctuation of speed and torque in the transmission system, which causes the non-working parts of the transmission to emit irregular reciprocating knocking noise within the allowable working clearance. Gear knocking noise mainly appears in the low-speed area during the acceleration of the hybrid acceleration mode, and its frequency spectrum presents a broadband characteristic. Gear whistling noise is a kind of noise caused by the dynamic excitation force of the internal gear in the meshing transmission, and the transmission error of the loaded gear in the meshing process. This kind of medium and high frequency pure tone is very easy to be recognized by the human ears, so it is The main NVH problem of gear transmission system[26].

There are many parts involved in enhancing the coupling stiffness of the transmission system, and the cost is relatively high. Therefore, without changing the gear material and heat treatment, making small changes to the gear and using the original design tool as much as possible, the target gear can be modified and optimized for the purpose of reducing the transmission error of the meshing gear, thereby reducing the noise of the meshing gear. Scholars at home and abroad have done a lot of research on gear modification technology. It is generally believed that gear modification is mainly divided into two categories[27]: tooth profile modification and axial modification. Liu[28] and YUAN et al.[29] use genetic algorithms to determine the tooth profile and tooth direction parameters that effectively reduce gear noise. Ghosh et al.[30] and Barlam et al.[31] study gear deformation and optimization of NVH performance through contour graphics and finite element slip line contact simulation models, respectively.

3.2.2. The NVH challenge of engine start-stop. The engine is mainly excited around the crankshaft at the moment of starting and stopping. For the purpose of torque and battery power and fuel saving, the engine on the HEV needs to be started and stopped frequently. The continuous switching of the start and stop modes of the engine can cause various transient shocks, as shown in Figure 5, and will be accompanied by high-frequency noise that causes obvious discomfort to passengers.

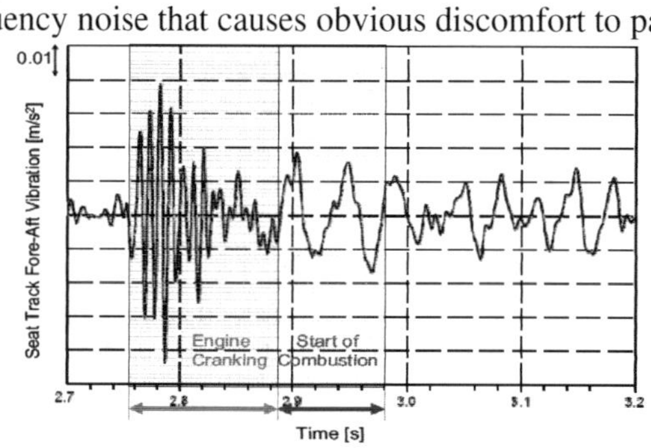

Figure 5. Vibration conditions when the engine starts[32]

In order to reduce the start-stop vibration impact, the current methods of electronic control technology are mainly implemented: adopting measures such as delaying the ignition advance angle, delaying the closing time of the intake valve, and controlling the fuel injection volume to improve the combustion of the engine; when the initial position of the control piston is after the intake valve closes and is close to the top dead center, the starting vibration amplitude is small; the starting torque of the generator acting as the starter is increased to reduce the torsional resonance of the power unit. Toyota's team[33][34] take their first and second generation hybrid systems as the research objects, and analyze the effects of various control methods on improving the NVH performance of the start-stop process. Aiming at the problem of unstable idling operation when the engine restarts, MENG[35] proposes the H-inf control algorithm to control the speed change. In addition, the suspension problem of the engine can be improved by shortening the distance between the power center of gravity and the rotating elastic shaft, and adding torsional vibration damping dampers. Sugimura et al.[36] and YUE et al.[37] improve the restraint on vibration and shock by adjusting the damping and stiffness layout of the transmission system.

3.2.3. The NVH challenge of HEV engine start-stop. Since the power source and corresponding transmission system composition of each working mode of HEV have to change, the characteristics of the torsional vibration problem of the transmission system will also change with the change of working mode.The quality of the dynamic coupling device is directly related to the efficiency of the dynamic coupling and the stability of the coupling process[38]. When the coupling device has problems in the process of synthesizing power, it will produce shock and vibration of the coupling device, and associated noise. Unsteady coupling power output can also cause vibrations in the power transmission system connected to it. In addition, the power coupling device is also responsible for the power decomposition and energy feedback functions, storing part or all of the engine power in the battery through the generator, which will cause the stability of the power switching process. The rapid change of working mode will cause the continuous switching and coupling of various torsional vibration problems with different characteristics to produce complex effects[39]. This aspect is summarized as the increase of HEV power sources and the switching of working modes profoundly change the inherent characteristics of torsional vibration of the transmission system, making it more complex and changeable.In order to achieve better fuel economy, the three types of HEV are equipped with supercharged engines with smaller displacement. The trend of engine miniaturization and supercharging has caused the problem of engine output vibration to be aggravated, and the torsional vibration problem of the drive train may be further exacerbated by the addition of fast-response motors. Due to the switching of HEV between various working modes, the coupling of different torsional vibration characteristics and the profound change of excitation response characteristics of power source lead to the complexity of natural vibration characteristics and excitation characteristics of transmission system. It is a new solution to suppress or reduce the torsional vibration of the transmission system based on the active control of the motor, rather than limited to the passive measures such as optimizing the structural parameters of the transmission system and installing matching damping devices. DAVIS[40], KIM[41] and ZHANG[42] track the torque fluctuations of the engine and clutch and optimize the control to reduce the influence of the switching process.

3.3. Vibration and Noise of Electrical Accessories System

Compared with traditional vehicles, HEV are equipped with electrical accessories such as electric air conditioners, electric vacuum pumps, and battery pack cooling systems. Since the noise of various electrical accessories is a factor that affects the NVH performance of hybrid electric vehicles, it is necessary to optimize the corresponding NVH performance of different electrical accessories.

3.3.1. NVH problems of battery pack cooling system. There are generally several ways to cool the battery pack: natural cooling, fan cooling and water cooling. The natural cooling method is difficult to cool the battery pack at low speed or idling speed, which is not only inefficient but also unreliable. Therefore, the latter two measures are mostly used for HEV[43]. The sound energy of the water pump and cooling fan is not too great when they work alone, but because the noise frequency is relatively high, and there is no shielding of engine noise, it is easy to be perceived by passengers. In power battery systems, fan cooling is more common. The fan cooling solution is superior to liquid cooling in terms of reliability and cost, but it needs to solve its complicated noise problem. Modifications around the fan's noise characteristics may lead to changes in aerodynamic performance such as static pressure and flow, and these performances must be guaranteed in actual engineering applications, and aerodynamic performance cannot be sacrificed for noise performance.Therefore, it is not recommended to modify the fan structure to reduce the noise of the fan. Fully consider the cooling requirements of the system and match the appropriate cooling fan before considering the noise characteristics of the fan.

The noise of the battery fan system mainly appears in the following aspects[44]: 1. The fan noise transmitted through the air; 2. The structure noise transmitted through the structure connecting the battery and the body; 3. The fan duct is caused by the vortex of the fan airflow noise. Therefore, for the noise of the battery cooling system, it is necessary to reduce the sound source noise. In addition to reducing the noise of the fan itself, it is also necessary to control the derivative noise generated by the fan, such as the noise generated by the battery components and the duct wall under the excitation of the fan. To reduce the noise of the fan itself, it is necessary to adopt multi-level speed control for the fan. By reading the information on the CAN bus and verifying with the battery temperature, the speed control strategy is implemented to ensure the speed of the battery fan under different working conditions. In the proper range, the battery temperature is also taken into consideration without exceeding the noise level. For HEV, the noise in the passenger compartment is minimal when the vehicle is stationary and the engine is not started. At this time, the high-speed start of the fan has the greatest impact on the noise of the occupants. Therefore, the key to the control of HEV fan noise is to keep the noise as low as possible under stationary and engine non-starting conditions. The structure of the battery cooling system can be optimized in terms of the cover plate of the air inlet and the thickness of the wall of the air outlet pipe. Singh et al.[45] and Scheit et al.[46] use the method of changing the shape and angle of the fan blade to reduce the fan noise of the cars.REN et al.[47][48] analyze the axial and radial vibration of the battery cooling fans in electric vehicles, and reduce the vibration energy in the axial and radial directions through improved schemes. WANG[44] suppresses the fan noise of HEV by adjusting the speed, changing the structure of the system, and improving thermal management.

3.3.2. Noise and vibration of air-conditioning system. The electric vehicle air conditioning system is mainly composed of electric compressor, condenser, evaporator, condensing fan, blower, connecting pipeline and other components. The noise sources during the operation of the air-conditioning system mainly include two categories[49]: 1. Fan and air duct noise: wind noise generated during the operation of the condenser fan and the air-conditioning box blower, and the mechanical vibration noise caused by the vibration and transmission of the fan; 2. Compressor vibration noise: including the radiated noise emitted by the compressor itself and the noise generated by the vibration of the structural parts caused by the compressor vibration.

The compressor is the power source of the refrigeration system and the main noise source.There are unbalanced inertial forces of the motor rotor and other rotating parts in the compressor, flapping of the valve plate, and friction between relatively moving parts. These factors cause the compressor itself to generate relatively large vibration and noise. At the same time, the vibration and noise of the compressor are transmitted to the body and the interior of the car through the structural connectors, which affects the subjective feelings of the drivers and passengers. The noise sources of compressors are complex and affect each other. So far there is no systematic method to conduct a comprehensive and systematic analysis of the noise sources of air-conditioning compressors and their control methods. Published papers and research reports generally only involve part of the excitation source or vibration isolation scheme. There are many separate studies on the mechanical noise, aerodynamic noise and electromagnetic noise of the compressor[50]-[53].

3.3.3. NVH characteristics of electric vacuum pumps. To ensure that there is enough vacuum source when the engine is not working, an electric vacuum pump is added to the brake booster system. According to the "incentive source-transmission path-receiver" model, NVH problems caused by the operation of electronic vacuum pumps mainly rely on two types of transmission paths, as shown in Figure 6: 1. Air transmission: The noise emitted by the electric vacuum pump when it is working is spread from the engine compartment to the atmosphere, and then passed into the cab through the air. This part of the noise is transmitted through the atmosphere and is lost more, especially when the windows and doors of the cars are completely closed, this part of the noise is basically not perceived and can be ignored. 2. Body transmission: The noise emitted by the electric vacuum pump is transmitted by the mounting bracket and the front panel of the body, and is sensed by the occupants after being amplified by the approximately closed cockpit. This part of the noise is the main noise perceived by the occupants.

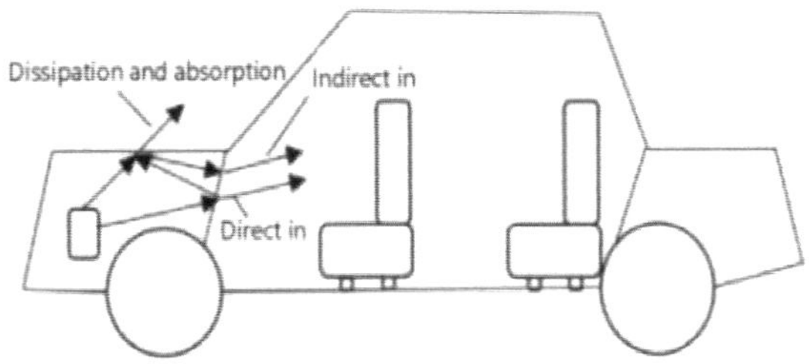

Figure 6. Vibration and noise transmission o vacuum pump[54]

Since the noise transmitted in the air has been weakened during the transmission process, and it cannot be improved from the transmission path, this part of the noise can be ignored in practical applications. The noise transmitted through the vehicle body can be reduced by changing the noise source, damping, isolating vibration, and changing the transmission path. NVH control for electric vacuum pumps mainly involves adjusting the installation position of the electric vacuum, designing a sound insulation structure in the cockpit, and diagnosing noise problems through experiments, and then combining engineering experience to optimize NVH performance. YU[55], GU[56], SONG[57] and LIAN[58] and others achieve NVH performance control by adjusting the installation position and working performance of the electric vacuum pump or arranging solutions such as vibration isolation, sound insulation and environmental sealing.

3.4. Wind Noise and Road Noise

The shielding effect of engine noise on HEV is weaker than that of traditional fuel vehicles. Therefore, the contribution of wind noise and road noise to the noise source of HEV has increased. In order to pursue high fuel economy, low-friction tires are usually used on HEV, and weight reduction is also being pursued in the design of body structure. These changing factors have brought challenges to the control of road noise and wind noise of HEV.

3.4.1. Types and solutions of road noise. The road noise of HEV can be divided into structural road noise and aerodynamic road noise. The excitation forces that cause structural road noise include: the vertical force generated by the continuous local compression and release of the contact surface of the tire and the road surface, and the longitudinal force generated by the continuous rolling and release of the tire and the road contact surface. The excitation force is transmitted to the axle through the tire and rim coupling system, and the axle is transmitted to the chassis and body[59]. The frequency range of structural road noise is generally between 20 and 400 Hz, which belongs to medium and low frequency noise. Aerodynamic road noise includes the noise generated by the compression and release of air in the interaction between tires and the road, and the noise generated by the friction between the tire and the road and transmitted to the car through the air. The frequency range of air road noise is 400-5000Hz. It belongs to medium and high frequency noise.

Based on the mechanism of road excitation noise, it can be controlled from excitation source, transmission path and response[60]:1. In terms of tire design, performance can be improved by optimizing the tire structure and material and tread pattern design; 2. Optimize the vibration isolation performance, structure and bushing parameters of the suspension to reduce the vibration transmission from the suspension to the body mounting point; 3. Vehicle body response control, including the stiffness of the suspension mounting point and the control of the vibration transfer function and noise transfer function into the vehicle, as well as the decoupling of the overall and local modes of the vehicle body. In the above-mentioned control method, the optimization of tire and suspension parameters sometimes conflicts with ride comfort and stability. Therefore, it is necessary to comprehensively consider all aspects of vehicle performance in the optimization process.

3.4.2. Types and solutions of wind noise. The As shown in Table 2, the wind noise generated during driving can be divided into four types[61].The energy of this kind of pulse brought by wind vibration noise to the human ears is very large, and the noise level can exceed 100dB, which makes the occupants feel uncomfortable. Buffeting noise occupies the largest proportion of vehicle interior wind noise, so the research work on it is currently the most. Combining the control of the wind noise source and the optimized design of the transmission path, the wind noise can be suppressed from the following perspectives[62]: ①The whole vehicle design: including the design between the overall flow line of the body and the engine compartment cover, the engine compartment and the front wind design between window glass; ②Partial design: design of reflector, design of sunroof; ③Dynamic sealing design: sealing between door and body, and sealing between glass and sealing strip.

Table 2. Types of wind noise and control methods

Types	Causes and characteristics	Control Methods
Leakage noise	Due to the existence of gaps, leakage noise appears in the car body and is transmitted to the car; mainly concentrated in the high frequency band, making a "whoop" sound.	Under the condition of ensuring that the static sealing of the body is not problematic, the dynamic sealing of the body is further improved to effectively reduce the leakage noise.
Cavity noise	The noise generated when the air blows to the gaps outside the HEV and the small cavity formed between the car body and the lap joint; the frequency is mainly determined by the air velocity and the size of the cavity passing through.	Shorten the gap between the outer panels of the car body; fill the cavity structure; and avoid the resonance sound generated after the airflow is blown into the gap and cavity.
Pulsating noise	The noise generated by pressure fluctuations on the surface of the car body is a broadband noise that is not easy to make occupants feel annoying.	Start with the overall body design and local design to reduce pressure pulsation and turbulence.
Buffeting noise	When the sunroof or window is opened during the driving of the car, a strong roar is generated in the car; the frequency is very low, and the occupants feel uncomfortable due to the feeling of pressing their ears.	Control the interaction between the airflow outside the car and the cavity inside the car; and design a reasonable sunroof spoiler.

3.4.3. Active noise control. At present, most of the control of wind noise and road noise still adopts passive control schemes, such as sound absorption and insulation, and adding acoustic packages, which contradict the trend of lightweight vehicles. In addition, using the above methods also bring great damage to the speech signal, and even eventually reduce the readability of the speech signal.
ANC (Active Noise Control) is an effective noise reduction method proposed in the early 20th century, which can reduce target low-frequency noise (usually below 500 Hz). The principle of ANC is to generate secondary destructive anti-noise signal with the same size and opposite phase as the engine and transmission system to suppress noise[63]. By reasonably arranging microphones and secondary sound sources in the vehicle cab to adaptively eliminate broadband random noise, it can effectively control mid-frequency and low-frequency noise, without changing the body structure through topology optimization, and without having to install heavy and high-cost sound-absorbing materials. By reasonably arranging microphones and secondary sound sources in the car, adaptively eliminating broadband random noise, and real-time feedback and optimization of the online control system, it can effectively adapt to various working conditions and reduce mid-frequency and low-frequency noise. There is no need to change the body structure through topology optimization, nor to install heavy and expensive sound-absorbing materials, so it is of great promotion significance[64][65].

4. Summary

A variety of power sources are used to drive the HEV, and its working conditions are diverse and complex. HEV often switches between stable operating conditions and transitional operating conditions, so its vibration and noise characteristics are much more complicated than traditional fuel vehicles and pure electric vehicles. The rapid development of HEV presents new challenges for NVH problems and brings new opportunities. For NVH problems that have similar vibration and noise characteristics to traditional fuel vehicles, existing vibration and noise reduction measures can be adopted. For some unique NVH problems of HEV, new prevention and control technologies should be actively researched to improve the NVH performance of HEV.

5. Reference

[1] Ajay Babu, S. Ashok. Energy and fuel efficient parallel mild hybrids for urban roads[J]. Energy Conversion and Management,2016,121.
[2] Yao L. The Parameter Matching and Optimzation of Power Source for Parallel Hybrid Electric Bus[D]. Beijing Institude of Technology, 2016.
[3] Albert A, Schwarz A. Automated Identification of NVH- Phenomena in Vehicles[C]// SAE 2011 Noise and Vibration Conference and Exhibition. 2011.
[4] Lan Y, ZUO Y.Application of Partial Coherence Analysis to Internal Noise Source Identification for Hybrid Electric Vehicles[J]. NOISE AND VIBRATION CONTROL, 2019, 39 (06) :89-94.
[5] Komada M, Yoshioka T. Noise and Vibration Reduction Technology in New Generation Hybrid Vehicle Development[C]// Sae Noise & Vibration Conference & Exhibition. 2005.
[6] Yu F. Torsional Vibration Analysis and Control of Transmission System of CVT Hybrid Electric Vehicle[D]. Jilin University,2019.
[7] Li X. Introduction to Electric Vehicles[M]. Beijing Institute of Technology Press, 2005.
[8] Xiong J, Huang J. Analysis and Control of Noise and Vibration of Hybrid Electric Vehicles [J]. NOISE AND VIBRATION CONTROL,2009,29(05):96-100.
[9] Sun F, He H. A Study on the Method of Classification of Hybrid Vehicles[J]. JOURNAL OF BEIJING INSTITUTE OF TECHNOLOGY,2002(01):40-44.
[10] Zhao H. Hybrid Electric Vehicle Technology[M]. Beijing: China Machine Press, 2012.
[11] Lin C, Wang S, Moallem M, et al. Analysis of Vibration in Permanent Magnet Synchronous Machines due to Variable Speed Drives[J]. IEEE Transactions on Energy Conversion, 2017, 32(2):582-590.
[12] Park S, Kim W, Kim S. A Numerical Prediction Model for Vibration and Noise of Axial Flux Motors[J]. IEEE Transactions on Industrial Electronics, 2014, 61(10):5757-5762.
[13] Zhu H. Analysis on electromagnetic exciting force of asynchronous motor[D]. Zhejiang University, 2013.
[14] Chen Y. Analysis and Control of Motor Noise[M]. Zhejiang University Press, 1987.
[15] Song Z. Research on axial-magnetic -field-modulated brushless double-rotormachine for hybrid electric vehicles[D]. Harbin Institute of Technology, 2017.
[16] Zheng G,Wang Z,Wang B.Dynamic Characteristics Analysis of Stator System of Permanent Magnet Synchronous Motor.[J].Journal of Chongqing University of Technology (Natural Science Edition),2019,33(6):53-58. DOI: 10.3969/j.issn.1674 -8425(z).2019.06.008.
[17] Salon S, Ergene L T. Design and construction of a low-vibration low-leakage field motor[C] // International Conference on Electrical Machines. IEEE, 2014: 464 -469.
[18] Wang X, He X, Gao P. Research on Electromagnetic Vibration and Noise Reduction Method of V Type Magnet Rotor Permanent Magnet Motor for Electric Vehicles [J]. Proceedings of the Chinese Society of Electrical Engineering, 2019, 39(16): 4919-4926+4994.
[19] Hur J, Reu J W, Kim B W, et al. Vibration Reduction of IPM-Type BLDC Motor Using Negative Third Harmonic Elimination Method of Air-Gap Flux Density[J]. IEEE Transactions on Industry Applications, 2011, 47(3):1300-1309.

[20] Chai J Y, Lin Y W. Comparative study of switching controls in vibration and acoustic noise reductions for switched reluctance motor[J]. IEE Proceedings-Electric Power Applications, 2006,153(3): 348-360.
[21] Liu H, Liu Q, Zhang W, Miao Y. Random PWM Technique for Acoustic Noise and Vibration Reduction in Induction Motors Used by Electric Vehicles[J]. Transactions of the China Electrotechnical Society, 2019,34(07):1488-1495.
[22] Gwak D Y, Yoon K, Seong Y, et al. Application of subharmonics for active sound design of electric vehicles[J]. Journal of the Acoustical Society of America, 2014, 136(6): EL391.
[23] Taniguchi S, Homma T, Wakao S, et al. Control Method for Harmonic Voltage Injection to Achieve Noise Reduction in Position-Sensorless Control of Permanent-Magnet Synchronous Motors at Low Speeds[J]. Electrical Engineering in Japan, 2012, 179(3):49-56.
[24] Jiri T. Gearbox Noise and Vibration Prediction and Control[J]. International Journal of Acoustics and Vibration，2009，14(2): l-11.
[25] Sun M, Lu C, Liu Z, et al. Classifying, Predicting, and Reducing Strategies of the Mesh Excitations of Gear Whine Noise: A Survey[J]. Shock and Vibration, 2020, 2020:1-20.
[26] Li Y. The study on Vibration and Noise Control of Automobile Gearbox[D]. Chongqing University, 2004.Amatucci G G, Tarascon J M, Klein L C. Cobalt dissolution in LiCoO2-based non-aqueous rechargeable batteries. Solid State Ionics, 1996, 83: 167–173
[27] Zou J. Gear Whine Noise Reduction of Electro-Mechanical Coupling Transmission Based on Gear Profile Modification [D]. South China University of Technology, 2018.
[28] Liu Z. Transmission noise control based on gear modification[D]. Jilin University, 2017.
[29] Yuan Z, Sun Z, Guo Y. Optimal Design of Tooth Profile Modification Curve of Straight Tooth Cylindrical Gear [J]. Mechanical Transmission, 2010, 34(5): 1-4
[30] Ghosh SS, Chakraborty G, On optimal tooth profile modification for reduction of vibration and noise in spur gear pairs[J]. Mechanism and Machine Theory, 2016, 105: 145-163
[31] Barlam D., Zahavi E.. The reliability of solutions in contact problems[J], Computers & Structures, 1999, (70): 35-45
[32] Integration in Hybrid Vehicles[J]. Sae International Journal of Passenger Cars Mechanical Systems, 2009, 2(1):1396-1405.
[33] Yoshioka T, Sugita H. Noise and vibration reduction technology in hybrid vehicle development[R]. SAE Technical Paper, No. 2001-01-1415, 2001.
[34] Tomura S, Ito Y, Kamichi K, etc. Development of vibration reduction motor control for series-parallel hybrid system[R]. SAE Technical Paper, No.2006-01-1125, 2006.
[35] Meng W. H-inf control research for engine idle start and stop system [D]. Harbin Institute of Technology, 2014: 27-46.
[36] Sugimura H, Takeda M, Takei M, etc. Development of HEV Engine Start-Shock Prediction Technique Combining Motor Generator System Control and Multi-Body Dynamics (MBD) Models[R]. SAE Technical Paper, No. 2013-01-2007, 2013.
[37] Yue Z, Liu Y, Xie K, et al. Noise & Vibration Characteristics and Control in Hybrid Electric Vehicle[J]. Automotive Technology,2018,(11):27-31.DOI:10.19620/j.cnki.1000-3703.20170637.
[38] Wang Y，Qin X，Huang S，et al. Design and analysis of a multi-stage torsional stiffness dual mass flywheel based on vibration control[J]. Applied Acoustics，2016，104: 172-181.
[39] Yan M . Analysis and control of torsional vibration of hybrid bus transmission system[D]. Tsinghua University，2015.
[40] DAVIS R I, LOVENS R D. Engine torque ripple cancellation with an intergrated starter alternator in a hybrid electric vehicle: implementation and control[J]. IEEE transactions on industry applications, 2003, 39(6) :1765-1774.
[41] KIM S, PARK J, HONG J, et al. Transient control strategy of hybrid electric vehicle during mode change [C] / / SAE Technical Paper. 2009.

[42] Zhang H , Wang C L , Zhang Y . Drivability improvements for a single-motor parallel hybrid electric vehicle using robust controls[J]. Journal of Zhejiang Universityence A, 2014, 15(4):291-301.
[43] Tan X. Design of Electric Vehicle Power Battery Management System [M]. Sun Yat-sen University Press, 2016.
[44] Wang L. Application and Research of Noise Control of battery fan noise in hybrid Automotive [D]. South China University of Technology, 2013.
[45] O P Singh，Rakesh Khilwani，T Sreenivasulu. Parametric study of centrifugal fan performance: experiments and numerical simulation[J].International Journal of Advances in Engineering & Technology , 2011, 1(2):33-50.
[46] C. Scheit，B. Karic，S. Becker.Effect of blade wrap angle on efficiency and noise of small radial fan impellers-A computational and experimental study.[J].Journal of Sound & Vibration，2012,331(5):996-1010.
[47] Ren J, Zhang X, Shi H. Analysis and Improvement of Axial Vibration of New Energy Vehicle Battery Cooling Fan[J]. Journal of Henan University of Science and Technology (Natural Science Edition), 2017, 38(4) : 25-29,34.
[48] Ren J, Zhang Xi, Hu B. Radial vibration test and reduction method on battery cooling fan of electric cars[J].Manufacturing Automation,2018,40(6):147-150.
[49] Luo Y, Xu W, Yu X. Vibration Noise Analysis of Air Conditioning System for Electric Vehicles[J]. Automotive Engineers, 2019(08): 41-44.
[50] Lu X, Yang C.Identifying Research on Noise Sources of Rotary Vane Compressor[J]. Journal of Chongqing University (Natural Science Edition), 2006(08): 70-73.
[51] Cuevas C, Winandy E, Lebrun J. Testing and modelling of an automotive wobble plate compressor[J]. International Journal of Refrigeration, 2008, 31(3):423-431.
[52] Zhang S. Optimization Design of Vibration Reduction for Air Conditioning Compressor and pipe[D]. Southwest Jiaotong University, 2015.
[53] Li S, Wang K, Xu W. Reseach on Noie and Vibration Reduction of Compress of Con ditioner[J]. Noise and Vibration Control, 2003(03): 27-30.
[54] Li D,Tang L, Zhao W,Hu J.Study on the Calculation Method of Vehicle Interior Noise Based on Vacuum Pump Bench[J].Automotive Science and Technology,2020(03):16-21.
[55] Yu Y, Zhang G, Tao T, Zhang C. Noise analysis and optimization of vehicle electronic vacuum pump matching[J]. Light Vehicle Technology, 2013(04): 19-22+5.
[56] Gu C, Dong J, Yang X. Selection and optimization of vacuum pump for an electric vehicle based on acoustic and vibration measurement[J]. Electronic Measurement Technology,2018, 41(05): 78-81.
[57] Song J. Research on the Noise of Electric Vacuum Pump in Electric Vehicles[J]. Times Auto, 2017(16): 74-76.
[58] Lian J, Zhang J, Yang L, Ye C. Mechanisms of Interior Noise caused by Brake Vacuum Pumps of Electric Vehicles[J]. Noise and Vibration Control, 2019, 39(02): 114- 117.
[59] LI T，BURDISSO R, SANDU C.Literature review of models on tire-pavement interaction noise[J].Journal of Sound and Vibration，2018，420：357-445.
[60] Zhao C, Xing Y, Wang H. Method to Improve Vehicle Noise Excitated With Some Road[J].Automotive Electrical Appliances, 2018,(7):30-33.
[61] Pang J. NVH Control in Automotive Bodies [M]. China Machine Press，2015：431.
[62] Liu D. The Research and Implementation of Automobile Wind Noise ANC System Based on FxLMS Algorithm [D]. Chongqing University of Posts and Telecommunications, 2019.
[63] Sahib M A , Streif S . Design of an active noise controller for reduction of tire/road interaction noise in environmentally friendly vehicles[J]. 2017:59-62.
[64] Yuan J, Liu D, Lu W. Design of Active Noise Control System Based on Vehicle Wind Noise[J]. Journal of Chongqing University of Technology (Natural Science Edition), 2019,33(10):9-15. DOI:10.3969/ j.issn.1674-8425(z).2019.10.002.

IOP Conf. Series: Materials Science and Engineering 1194 (2021) 012009 doi:10.1088/1757-899X/1194/1/012009

[65] Ford Global Technology Company. Wind noise fluctuation reduction system and method: CN201810035540.1[P].2018-07-27.

AUTHOR INDEX

Institute of Physics
Dirac House, Temple Back
Bristol BS1 6BE UK

ISSN: 1757-8981
ISBN 978-1-7138-4653-6

CURRAN ASSOCIATES INC.
proceedings.com